D0041423

"What a strange heartbreak to fall in love with a girl dead for twenty-eight years. The U.S. government tried its best to kill off the memory of Deborah Gardner, but no such luck. Philip Weiss's portrait of her life and death is unforgettable."

—Richard Ben Cramer, author of
Joe DiMaggio: The Hero's Life and *How Israel Lost*

"A masterful job. . . . Weiss is succinct and unsparing."

—*Sunday Oregonian*

"*American Taboo* is not a whodunit but a how-did-he-get-away-with-it and a why-didn't-we-know-about-this. It ought to be an American legend. Now, thanks to Philip Weiss's brilliant narrative, it will be."

—Michael Kinsley, founding editor of *Slate*

"*American Taboo* is a kind of love story, the love of a writer for an idea, for the truth."

—*Times-Picayune* (New Orleans)

"A riveting narrative and an unforgettable portrait of American volunteers and Tongan officials. For those who've been tormented by this incident, Weiss's exhaustive work must now appear as a kind of catharsis. . . . There is a humbling and timely lesson here for all of us citizens of the last superpower."

—*Mother Jones*

"Filled with intrigue. . . . Weiss strongly conveys through researched detail a heartbreaking tragedy, a disgusting cover-up, and a reminder that justice can't be taken for granted."

—*Hartford Courant*

"Meticulously deconstructed. . . . One of the most exotic true crime books of recent years, and one of the saddest."

—*Washington Post*

AMERICAN
TABOO

ALSO BY PHILIP WEISS

Cock-a-Doodle-Doo

AMERICAN
TABOO

A Murder

in the

Peace Corps

Philip Weiss

HARPER  PERENNIAL

HARPER ⬤ PERENNIAL

Cover photograph of Emile Hons, Deb Gardner, and Rich Dann at Fua'amotu airport in Tonga, September 25, 1976, courtesy of Emile Hons.

A hardcover edition of this book was published in 2004 by HarperCollins Publishers.

P.S.™ is a trademark of HarperCollins Publishers.

HarperCollins books may be purchased for educational, business, or sales promotional use. For information please write: Special Markets Department, Harper-Collins Publishers, 10 East 53rd Street, New York, NY 10022.

First Harper Perennial edition published 2005.

Designed by Nancy Singer Olaguera

Library of Congress Cataloging-in-Publication Data is available upon request.

ISBN-10: 0-06-009687-X (pbk.)
ISBN-13: 978-0-06-009687-8 (pbk.)

05 06 07 08 09 ❖/RRD 10 9 8 7 6 5 4 3 2 1

FOR ELLEN AND LEON WEISS

Contents

AMERICAN
TABOO

1

A Legend of the South Seas

No one forgets his first foreign country. The light, the architecture, the way they do their eggs. Red money. The dreamy disorientation. The smell of aviation fuel.

I didn't choose Samoa, John did. We were both 22 and starting out on a long backpacking trip, and he bought tickets in Los Angeles with six stops down through the Pacific. Samoa was after Hawaii. We got there in January 1978. We stayed at a Mormon family's house in the capital, Apia, climbed through jungle to Robert Louis Stevenson's grave, then set out for the bigger western island. The Peace Corps volunteer was on the ferry, a redheaded guy with half a Samoan marriage tattoo on his back. Of course it turned out Bruce and John had grown up a few miles away from one another in Montgomery County, Pennsylvania, so he had us back to his seaside village. We met his Samoan wife Ruta and stayed two or three nights.

It rains harder in Samoa than anywhere. The rain against the metal roof made a throaty song that rose and fell, and under a kerosene lantern, as we dined on one of his chickens, Bruce told us about the murder.

A year or so back in the neighboring country of Tonga, a male Peace Corps volunteer had brutally killed a female volunteer by repeatedly stabbing her. There had been some kind of triangle, a Tongan man was involved. Then the American man was gotten off the island. The case had caused all kinds of tension between the Peace Corps and island governments.

Bruce didn't know more than that, didn't know names or dates. The story had passed from one island to another as stories always did in Polynesia, by word of mouth. The only difference between this story and others was that it involved Americans.

And already then, when I heard the legend in my first foreign country, there was a sense that something was wrong. That the original wrong had been compounded.

Ten years went by. I started working as a journalist in New York, and one night at a bar I met another writer, who said that he had been in the Peace Corps. "Where?" "Tonga. I was in Tonga 1, the first group of volunteers to the Kingdom."

I asked whether he had heard Bruce's story.

"Oh, yes," Fred said. "Later volunteers told me something. Elsa Mae Swenson, that name comes back. That was the victim."

Her name was Deborah Ann Gardner. The next day in the New York Public Library I found the one article about the case that appeared in the *New York Times*, an inch or two at the bottom of page 7 in January 1977. The wire story was based on an account from the *Chronicle*, the government newspaper of Tonga, and said that the male volunteer was from New York and a Tongan jury had found him to be insane when he killed her.

Of course I looked him up in the New York phone book, and there he was. He had been listed from a couple of years after the murder.

I called the Peace Corps. Privacy law would be an important factor in any disclosure. "His rights are basically uppermost at this time," a lawyer explained. So I made a formal request under the Freedom of Information Act, and a few months later a package of old records arrived at my apartment with a lot of the pages blacked out.

Deborah Gardner was 23 and a teacher. She lived in a one-room hut in a village at the edge of the Tongan capital of Nuku'alofa. She had been there nearly ten and a half months when she died in October 1976. The older volunteer charged with her murder faced possible hanging. The American government went to considerable lengths to defend him. A lawyer came from New Zealand and a psychiatrist from Hawaii. It was the longest trial in Tongan memory.

After the insanity verdict, the two governments went back and forth. Then the King of Tonga and his cabinet released the man on written assurance from the Americans that he was to be hospitalized back in the United States.

He refused to enter a hospital. The Peace Corps had lacked the power to make him do so, or the will. The case quietly disappeared.

The key was Deborah Gardner's family. Why had they never come forward? Their names and addresses were blacked out of the file on privacy grounds, and though she was from the Tacoma area, there were hundreds of Gardners listed in the local phone books. I made a few calls and sent a few letters, but before long I got on to something else and, telling myself I would return to this story someday, I put the file away in its big rough brown envelope, put the envelope in a box, and put the box in the attic.

Someday turned out to be 1997. I was hiking with a writer friend when he said that *Travel and Leisure* magazine was sending him to, of all places, the Kingdom of Tonga because it would be the first country in the world to see sunrise on the millennium and had announced a giant celebration.

"That's funny, I have a Tonga story," I said, and told him about the murder.

Michael stopped in the path. "Why are you working on anything else?"

I dug out the old file and searched for any clues to the identity of Deborah Gardner's family. A fellow volunteer had accompanied her body home. Though the name of the "boy escort" was blacked out on privacy grounds, some of the blackouts were sloppy and it was possible to piece his identity together. Emile Hons of California.

An Emile Hons was listed in San Bruno. I called a few times and left messages, finally got him.

Yes, he'd been in Peace Corps/Tonga. Now he ran the big shopping mall in San Bruno. He was guarded, and questioned my information.

"There may have been a triangle, but if there was it involved two *pālangis*, not a Tongan," he said, using the Polynesian word for a person of European descent.

We chatted and Emile seemed to get more comfortable.

"I lived right next to Debbie. I went to the movies that night. When I got home, I opened her door and it was solid blood. I was lucky to bring the body home. She was a beautiful person. Physically, but in spirit also. She was very kind."

But he hadn't stayed in touch with the family, didn't remember their names.

"Her parents were divorced. She had a big woodsy-type guy as a father."

I called Peace Corps headquarters again to see whether the long passage of time might have eased restrictions on the family's names. The information officer was quite surprised.

"This happened twenty-one years ago," he said. "You're the second person to call me about it today."

Jan Worth was a former Peace Corps volunteer, now a poet in Michigan. She was writing a novel about her Tongan experience with the murder in the background. She still thought about Deborah Gardner.

"I'd just arrived in the country, and Peace Corps had a big party for us. We were so grubby, and we were supposed to be modest. She showed up in this white dress. Long brown hair. Beautiful skin. I thought, How dare she look so great? Then a few days later she was dead. Part of the power of the event was her beauty. You couldn't talk about her without there being this halo over her.

"Then all of a sudden we were cast in the dark shadowy side of things. That shadow lasted for the whole time we were there."

Jan had put a classified ad in a Peace Corps newsletter, seeking information about the murder, and several former volunteers had gotten in touch with her to say they were still troubled by the case. One of them had also begun to write about it. So now there were three of us trying to get it into print, trying to end the legend.

For Mike Basile, the murder was the most tragic experience of his life. He had been number two on the Peace Corps staff in Tonga. Now in his mid-fifties, an academic in international studies, he felt the need to make a public record of these events before it was too late. He had begun to do so in the form of long letters to a Peace Corps buddy from his own service in Turkey.

In one letter, Basile described going with his boss to the Tongan hospital at 1 A.M. on the night of the murder.

"Dr. Puloka and Mary, my director, were sitting in the air-conditioned morgue beside the bed on which Debbie's body lay. Puloka rose when we entered, lifted the linen that had been placed over her naked body, her face remaining uncovered throughout. 'She showed some signs of life,' he reported, with some emotion, 'shortly before she was brought here by her neighbors. . . .' I looked, fixed on her. She was beautiful. The puncture wounds were not so evident to my furtive eye, as I concentrated on her face, one that I had seen in my office only the previous week. . . .

"She was sensual, full of young womanliness."

Mike had used Deborah's real name, but was calling the murderer "Joseph." In her novel, Jan was calling him "Mort Jensen."

Both were apprehensive. Deborah Gardner had been stabbed twenty-two times with a large knife, apparently out of feelings of jealousy and betrayal, then the man had gone free and no one knew about it. One reason the case haunted people was their sense that the killer had controlled its outcome, manipulating a number of other parties with a strong interest in the matter, and that the U.S. government—or some small segment of it, anyway—had gone along if only because it did not want the story to come out.

Jan Worth and Mike Basile both understood themselves to be breaking an old seal just by sitting down to write about it. "Something that scares me, twenty-one years later, could people that want to talk about it be targets?" Jan said. "Do you think the story will finally be told?" Mike said.

I told Jan that when she finished her novel I'd write an article about it— First Novel Unearths Legend Of 1976 Peace Corps Murder—and go knock on the killer's door.

So I called him once, just to hear the sound of his household on a Saturday morning. A man answered with a somewhat wired, ethnic voice, sounding alone. I hung up.

A few seconds later my own phone rang. I looked at it as though it might hurt me, and didn't touch it.

A year or two later, Jan sent me the first chapter of her book.

"For most of the volunteers, coming to the Kingdom was just about the sexiest thing they'd ever done," she wrote. "You could feel the heat in any group of volunteers. They found each other in dark corners and at the end of the table at the Tonga Club, and talked passionately about everything that was going on. Usually somebody'd be feeling somebody else's thigh at the time. The volunteers eventually got together when the desire—as physical as thirst and not at all discriminating—was just too much to bear."

I'd thought about this story half my life and never really left my desk. I made plans to go to Tacoma.

It was a cool, sunny day in April 2000, one of those rare bright days where everywhere you go around Puget Sound, you can see Mount Rainier hovering at your shoulder. I parked on Eleventh Street downtown and walked uphill to the Tacoma Public Library.

A librarian had informed me by telephone that the *Tacoma News Tribune* had printed nine articles about Deborah Gardner in 1976 and 1977. The newspaper's index and back issues were to be found at the library.

The articles were a sharp disappointment. They were all short wire accounts, most of them from the trial in Nuku'alofa in December 1976, and I'd read all but one or two before in the Peace Corps file. The headlines identified Deborah Gardner as a "Steilacoom Woman," because she was from the suburban town of that name. But no one at her hometown paper had bothered to find out anything about who she was. I'd hoped to see her picture. The paper never printed one.

I started to put my things away when Brian Kamens, a librarian, stopped at my desk and wordlessly set down a 3-by-5 index card.

"Gardner, Deborah A. 10-18-76," was typed across the top, and under that was pasted a small yellowing obituary.

"Deborah A. Gardner, 23, a Peace Corps volunteer from Steilacoom, died Thursday in Tonga. . . ."

Here were the family's names. Deborah's mother, Alice, lived at 2822 Garden Court, Steilacoom. Her father Wayne lived in Anchorage, Alaska. Her brother, Craig, lived in Pullman, Washington.

The obituary had appeared in the newspaper that I'd just searched, but the *News Tribune* had failed to include it in the index. Kamens had had the good sense to search a large card file of local obituaries that a collections agency had created and donated to the library some years before.

Why hadn't the obituary appeared in the index? The date, 10–18–76, loomed as the most significant thing about it. Deborah Gardner was not to become an international news story for another two weeks, 11–2–76.

I got the Peace Corps papers from my knapsack to figure out the sequence.

October 14, Deborah dies in Tonga.

October 16, Emile Hons brings her body home.

October 18, the obituary appears, offering not the slightest indication that Deborah was murdered. A funeral is held.

November 2, nineteen days after the murder, the Peace Corps issues an oblique statement in Washington. One volunteer has been slain, another charged with murder in Tonga.

The timing seemed underhanded. November 2, 1976, was Election Day. The papers were filled with news of Jimmy Carter's victory and other election results. Little wonder the case had fallen through the cracks, little wonder the *Washington Post* and *New York Times* had no room for Deborah Gardner. (The one little *Times* story did not appear for another three months.) I thought of the Peace Corps lawyer's comment, "His rights are basically uppermost at this time," and felt pity for Deborah Gardner.

Kamens was back, this time with a fat book. The obituary said that Deborah Gardner had attended Washington State University in Pullman, so he had found *The Chinook*, the WSU yearbook for 1974, which he opened on a picture of the women's Senate.

Deborah was womanly, relaxed, alive. She had big dark eyes, a strong jaw and stronger cheekbones, dark glossy hair, wide shoulders. She sat in the middle of a group of women.

In the middle, but somehow also apart.

The years fell away. Openness, informality, and independence flowed off her face, and now I was included in the circle that knew that this tragedy defied boundaries. Many years ago, Deborah Gardner traveled thousands of miles to serve her country, and what had happened to her in that faraway place? What had her country done in the aftermath, and why had it never come out? Suddenly these were matters of great urgency. Thanking Kamens, I got up to look for 2822 Garden Court.

2

Tonga 16

The old hotel was what you would expect of the government. Mossy and comfortable and narrow, neither too good or too bad, it perched over the Tenderloin on Taylor Street. Just up the hill was old San Francisco. Just down the hill was the city's underbelly. Prostitutes, junkies, panhandlers. The tall bony kid from Kansas saw someone being beaten up right on the street corner and reflected for the umpteenth time that day that he had never really been out of Kansas.

It was December 1, 1975, and one by one they came into the lobby, thirty-three in all. They carried red checkered rucksacks and external frame backpacks and duffel bags that their older brother had taken out to Vietnam and back, and at the reception desk they were given rooms on the eighth floor and told to meet at one o'clock in the mezzanine.

It was awkward at first. The East Coast people regarded the West Coast people warily; they seemed taller, quieter. A few clustered at the list of names taped to the wall and figured out the male-female ratio, 18 to 15, or silently counted the ethnic names, nine or ten. The tall bony kid from Kansas, David Scharnhorst, had made friends with Vic Casale, a fine-boned man from the east, and they sat together murmuring about the women. "She's pretty," said Scharnhorst. "That tall one? She's a horse," Vic said, and Scharnhorst felt corrected.

An official had come out from Washington to lead them. Dottie Rayburn was blonde with long fingernails, but easygoing. She had them sit in a big circle.

The purpose of these three days was to weed themselves out if it didn't sound right. And for those who were going it was to begin to familiarize themselves with the destination and use this time to buy everything they didn't have from the checklist.

They would get shots tomorrow and their passports Wednesday before the bus took them to the plane. Now it would be good if they all introduced themselves and said where they were from and why they were going. So they went around the circle speaking for about a minute each, and then an older volunteer got up to address them, a tall, dignified man who had been in the country two years.

"You are about to have the adventure of your life—this is something—you have no idea what you're getting into—but something that for me has been the greatest thing I've done," he said. "You are going to have some really incredible moments of being outside the United States and seeing America in a new way."

The door opened. A small man with lank blond hair had come in late and tried to close the door quietly behind him. Dottie turned around.

"Tonga 16?"

"Yeah, Tonga 16."

That was their name. Two or three groups a year had been going out to the Kingdom over the last eight years, and they were the sixteenth group.

Dottie told him to introduce himself and say why he was going, and he glanced around the room with pale silvery blue eyes. "My name is Francis G. Lundy, and I'm from Nacogdoches, Texas. I'm going out to teach physics and I've always wanted to see the constellations in the southern skies."

His voice twanged like a banjo, and a laugh moved through the room, the feeling that they all might share in his desire.

That night four of them went out to a Chinese restaurant, a little hole in the wall that Vic had found. He was the most charismatic of the volunteers, handsome, with a dark mustache and a nervous air, and a Graham Greene book in his back pocket. He had promptly renamed Scharnhorst Wichita, and walking down the street had examined Wichita's new knife to see whether he had done well, a long blade per the checklist, and he said that Wichita had done fine.

Then they came through the door and the restaurant was smoky and jammed. "I believe we're the only Occidentals here," Wichita said.

They got beers, and Vic leaned into the group.

"All this stuff about you got to weed yourself out, it's a mind game," he said. "They're just trying to get inside your head. It's government propaganda. They're just throwing a gauntlet. What they're trying to do is make you stay."

Wichita said it had been decided for him when he was 6 years old and Kennedy had driven past his door during the 1960 campaign. While the smaller fellow across from him, a good-natured outdoorsy guy from Colorado, Kelly Downum, said he had only decided at the campus rathskellar in May. He had broken up with a longtime girlfriend and didn't know what else to do. He had had too much to drink on a Friday, and someone came through with a clipboard. It felt like the French foreign legion to him.

Deb Gardner echoed that. She had broken up with a guy at Washington State and was sick of school and sick of Washington and ready for an adventure. "And if I can help some other people, why not," she said. She could teach science.

Back in the mezzanine, Vic had noticed her sharp cheekbones and thick gleaming hair, and glided toward her.

"It's Deborah but I go by Deb," she said.

"Well I'm Victor but it's Vic. I'm from New York."

"I'm from Pullman," she said, standing a little apart from him in her boots and jeans, unbeguiled. But just Pullman, as if Pullman were the center of Western civilization and not some cowtown in eastern Washington.

Now at the restaurant Vic said, "I feel like we're all in this together, we're going," and they drank to that. They were the four amigos, it was the foreign legion, and they were going to stick together through thick and thin.

The next day Tonga 16 got a bus to a veterans hospital for typhoid and smallpox shots, and back at the Hotel Californian there was a lot more about the country. It was very small, 95,000 people or so. There were some 40 populated islands but not a traffic light on any of them.

This was their last chance. If they were not ready for a slow pace and a strict

Christian culture, they should do themselves and Peace Corps a favor, and not get on the plane. If they were expecting bare-chested beauties pulling mangos off the trees, yes, there were mangos but the missionaries had covered the people up and volunteers were expected to work hard. The work could be frustrating. If they were looking for action, they should stay home and go to the movies.

Otherwise they would get their passports tomorrow and it was four flights across the Pacific from here, then two years. Two years of poverty and living as the people did, making 2 pa'anga a day, 3 American dollars, and teaching high school.

So it was their last night in the United States, and Francis Lundy walked down Nob Hill saying to himself, "I have cut my ties, I've cut my ties, I am completely free." He had not even told his mother back in Palestine, Texas, what he was doing. While Kelly made good on his determination to spend all his money before he left, and got so drunk that he came flopping down the eighth-floor corridor late that night, crawling, not able to walk, and calling out, "Peace Corps women! Peace Corps women!" And for some reason none of the women old or young answered his call.

The next day Dottie Rayburn handed out the no-fee passports, which expired in three years, and taking the passports was like stepping over a threshold.

Meet at four o'clock in the lobby. Did the women have long dresses and nylon underwear? Oh, and it would be culturally sensitive for the men to shave their beards.

Wichita was prepared to do anything for John Kennedy. He had already gotten his wisdom teeth pulled back in Kansas so as to minimize his chances of experiencing any dental problems over two years, and now he used his brand-new Buck knife and the shaving soap that was also on the checklist to hack off his beard in the hotel sink. Then when he came downstairs with his giant duffel bag no one recognized him.

A lot of them were smashed. Judy Chovan from Akron drank hard liquor for the first time in her life, and when they got to the airport at dusk she beat all the men at Pong. They boarded and Deb snuck up the stairs of the 747 to check out first class, then walked back down to the big group of volunteers at the back and smiled at the new family. People were passing around one guy's hiking boot he'd taken off, marveling at the size of it, 13.

Till there came a sharp cry and people looked around to see hip, quiet, Dave Johnson from Los Angeles going apeshit.

"My trunk's there on the ground. Our whole life is in that trunk."

Johnson had built it out of space-age honeycomb aluminum to hold as much weight as the government allowed him and his wife, Kay, including the finest stereo equipment anywhere in the world and a lot of cassettes. Before he screwed the lid down he had salted the contents with acid, barrels of blue microdot. "If you want to do drugs, do yourself and us a favor and don't come," they had been told. But the place shut down on Sundays, and how many traditional Christian Polynesians wearing woven mats around their waists would be the wiser if an American walked by out of his gourd?

"I'm going to call my congressman—"

Still the plane rolled out implacably from the gate, and stretched out the sunset going west, to Honolulu.

The group began to solidify. The stewardesses poured liquor all the way, and the volunteers shared tapes. The West Coast kids were into Santana, the Doobie Brothers, Jackson Browne. The East Coast kids had more black music, R&B, the Isleys, reggae, urban stuff, and English stuff. A serious guy from Connecticut had a meeting of the minds with a goofy guy from southern California over the latest release from the Dead, how the first cut went seamlessly into the second.

Roll away the dew
Roll away the dew

They got into Honolulu at nearly ten o'clock that night and laid over for a couple of hours, and though it was three in the morning on the East Coast, a girl from the Blue Mountains of Pennsylvania called her mother and woke her up, just to say goodbye.

Their second flight went overnight to the South Pacific, Pago Pago, and Deb sat with Vic at the back of the plane. He was three years older than she was, he'd been all over the world already.

"I figure it's going to be camping and reading," Deb said. "I've got a poncho, canteens, a tube tent, two backpacks." She'd thought of bringing the bike she'd taken around Europe with her brother but decided against it, she could buy one there. She'd brought books: Michener, Heinlein, microbiology texts.

"You're like some all-American chick, aren't you? Camping and biking. What were you, like a cheerleader?"

She shook her head, not giving him the time of day, but he had a soft voice, getting under her skin.

"Hey, I went out with a cheerleader—I was a quarterback in high school."

"Drill team," she said.

"Drill team—"

"Marching and dancing. You wore cool high white boots at the football games and had to get up early to practice."

"Did you bring the boots?"

Already he was teasing her. He said that she was hokey-pokey, a small-town girl, while he liked the big city and knew a couple of guys that were in. "In. You know. Connected. For an Italian, that means the Mafia," he explained.

And she countered with her own American vapor trail, she was part Indian. "Which you'd think would help me get into medical school, but no . . ."

"You're like half Cherokee, half Pepsodent."

"More like an eighth. My dad always says my great-grandfather was a tepee-creeper."

He asked about her father. He's in Alaska, she said. Was he a hunter? Yes, though he sold insurance and had remarried a woman who worked for an airline. That's how she'd gotten to Europe. Her father believed the Beatles had ruined America by showing up with long hair. But when Vic asked her more, she surprised him by starting to cry. She wasn't talking to her father.

"Don't talk to Papa?"

"No."

"That's too bad," Vic said mildly, but he had never heard of that, Vic whose father, Fiore Casale, anytime he went anywhere, said, "There is the front door, Victor, it is always open, you know that."

Then the stewardesses came around in the dark bearing bottles of champagne and paper cups. They had crossed the Equator.

"Did you notice how the farther we get, the planes keep getting smaller and smaller?" Wichita said.

They had switched to a prop plane for the short hop from Pago Pago to Western Samoa. A bus made from a flatbed truck took them into Apia for the night, and the houses that the bus passed, called *fales*, did not have walls. The government was treating them by putting them up at Aggie Grey's. Aggie's is the most celebrated hotel in the South Pacific, and the volunteers could not believe what they had stumbled into, thatched bungalows by the swimming pool named after Marlon Brando and Robert Morley, and Aggie's running waiters bringing lobster and steak while girls did the *siva*, a bewitching dance more hands than feet.

After dinner a group went out for American dancing. Vic led them down along the water to a nightclub with a mirror ball and a strobe, and the place was so hot their shirts stuck to the chairs when they jumped up to dance. They had bottle after oversized bottle of Steinlager, skunky New Zealand beer that they would soon be getting accustomed to, and didn't get back to Aggie's till after midnight.

The pool was dark, just a hair of moonlight.

"Who's going in?" Vic said.

"That's all right," said the others, and "Naa, we have to be up at five-thirty," and they drifted back to their rooms, and in the end only the two freest spirits went in, Vic and Deb. He pointed out the Southern Cross to her, though it was his first time seeing it too, and they tried to keep their voices down, tried not to splash.

The morning was sunny. A couple of men pressed their cameras to the airplane windows, taking pictures of the tooled blue ocean surface, when someone said, We crossed the international date line, and they went from Friday December 5, 1975, to Saturday December 6. Then another person called out that they had seen land. But just a clod of land, like some sort of oversight. Then there were others.

Tonga at last. How many had thought at first it was Africa, that Peace Corps had made a typo? Till one guy's surfer brother said, "No, it's the South Pacific and you'd be crazy not to go." Till Wichita got home from the university in Lawrence to discover that his fire-captain father had read the letter wrong, that he wasn't going to Togo.

Now here it was, the smallest kingdom in the world. It was hard to believe that this was a country, not just so many absentminded scatterings of sand and knitted green palms amid the sea's endless dark mood. Tonga's land mass was a quarter the size of the smallest American state, though it was bounded on the east by two great invisible things, the international date line and the Tonga Trench.

The propellers changed tempo from whine to moan, and they were descending toward a jawbone shape in the sea. Tongatapu was the Kingdom's biggest island, but not very big, about the size of Martha's Vineyard, and just as flat.

A welcoming committee of older volunteers had come out to the airport. One was particularly jovial, a gangling man in a white skirt with a big nose and a big beard, warm brown eyes, some kind of hippie guru making jokes a mile a minute.

"So they made you cut off your beards! They made us do that too, and boy did it piss me off when I got here and saw the Tongan cargo guys with beards pulling the luggage off the plane."

He introduced himself as Emile and filled them in on their new abode in a rush of charming patter.

"There isn't a traffic light in all the Kingdom and just three movie theaters. I call them the Pookay 1, Pookay 2, and Pookay 3." Pookay, spelled *puke*, meant sick in Tongan. "I've gotten sick in every one of them.

"You'll meet the King at New Year's. Just be sure and bow and call him Your Royal Heinie."

They sagged with laughter. It felt as if they'd been up three days straight. Then red leis were draped around their bleary necks and they were on a white coral road in a bus for the half-hour trip to two guesthouses in the city. "Oh it's so beautiful, it's so beautiful," one volunteer kept calling out in too breathy a voice. But others said to themselves, What is she seeing?

3

Swearing In

To an American, it looked like no city at all but like some town in the West that appeared after forty or fifty miles of back roads. One road was paved, the King's Road; the others exhaled pale clouds of coral dust when one of the few cars went by. Dogs chased pigs into the road, and back behind the mango and hibiscus trees, traditional *fales* could be seen, oval houses made all of coconut, with coconut posts and coconut thatch and woven coconut walls. Downtown only a handful of buildings rose two stories. Tungī Arcade was a modern structure where the *Chronicle*, the weekly newspaper, was prepared. Two blocks closer to the sea was the steepled Stone House, where the prime minister worked, and where the *Chronicle* was censored.

The only physical evidence of the American presence in the Kingdom stood on concrete stilts on the lagoon shore, a white barn of a building with an American flag painted on the clapboard, along with the words *Kau Ngāue 'Ofa*. This was the Tongan translation for Peace Corps: "They Work for Love."

Their trainers felt it essential that the new volunteers begin to learn the language away from the expatriate community in the capital so they got a boat to a neighboring island. The agricultural school they took over looked like a prison, and it rained for days. Everyone's clothes were wet. A tall goofy volunteer from southern California who looked like Eric Clapton walked around, chanting, "There is no vanity in Tonga, there is no vanity in Tonga," sometimes with a gecko poking out of his mouth. At least he was adjusting. The East Coast volunteers were in culture shock. Vic and the lanky academic guy from Connecticut stood together under palm trees, chanting the opening lines of *Gravity's Rainbow* at one another.

> *A screaming comes across the sky—*
> *It has happened before, but there is nothing to compare it to now.*
> *It is too late. . . .*

Everyone was thrown by the strange customs. You were not to expose the soles of your feet to anyone, that was the height of rudeness. You were never to touch anyone on the head, a Tongan could actually hurt you if you did that. There was no privacy. A Tongan trainer came up to a group of men to ask them if they had seen Francis's penis. They hemmed and hawed. Well, no, actually, they hadn't. He had seen it in the shower. "I really admired his penis. He has a very large penis."

They went to a beach in the rain, and a trainer held up a jar with what looked like a black potato inside. "This is a stonefish." It was one of the most poisonous animals in the world. If you stepped on its dorsal fin you might take a few steps down the beach, but then you would keel over. Well, who would ever go into the water after that? Rats came into their rooms at night and climbed the mosquito nets, and a lot of them got sick because the kitchen conditions were primitive. The Pennsylvania woman's smallpox vaccination blew up like a grape, and Kelly got so feverish they had to put him in a sleeping bag. It was 85 degrees outside but there he was, chilled and raving in a sleeping bag.

Then they understood what the older volunteers had told them back in San Francisco, they needed one another, Peace Corps couldn't take care of them. The Peace Corps doctor was stuck back in Nuku'alofa. He finally managed to get out to them by plane, then he walked around in the dark with a flashlight in one hand and a beer in the other, making housecalls. The doctor had long blond hair and looked like a surfer. Wichita had developed some kind of abscess from having his wisdom teeth out, and the doctor ran the flashlight over his face and grinned. "He looks like a chipmunk, doesn't he?"

Wichita did not see the levity in that. The realization had come to him that he'd given up a stipend in biochemistry at graduate school in order to come here and study a language that, tops, 200,000 people in the world used. He grew so sick of the rain that he put his chair out in it and let rain drip off his nose. Deb came out to visit him. "Faka'ofa 'a Wichita," she said. "You poor thing," and tilted toward him to hear everything that was wrong.

Then the neckline of her soggy red sundress gaped open and she was wearing nothing underneath and Wichita reached out chivalrously to paste it back against her collarbone.

"Deb, it's hard enough as it is," he said, and she gave him a sisterly smile.

Finally the skies cleared. The whole group went outside that night, and Francis Lundy gave a primer on the southern constellations. He pointed out the Southern Cross and the Magellanic Clouds, but he said that the southern constellations were not so bright and clearly defined as the northern ones. They were messier, there were few stark landmarks to compare to the Big Dipper or the W in Cassiopeia.

"But see, there's Orion—"

And they oohed excitedly to find Orion, but upside down.

Every New Year's Eve at midnight, the King held a reception at the Royal Palace to thank the pālangis who were serving the Kingdom, and on the last night of 1975, the new volunteers, having returned to Nuku'alofa, met up on the south side of the city.

Deb and Vic and Kelly started out at 8 from their guest house. They walked a half mile in the dark along pitted unmarked roads till the focus of expatriate life announced itself with a square of yellow light against the bush. The Tonga Club was a sagging low-slung shack with a dilapidated banana tree on either side of the front door. Tongan policemen and clerks and radiomen ran the club, but pālangi volunteers were honorary members. Wichita sat at a formica table nursing a boilermaker under faded portraits of the King and his late mother, Queen Sālote, that hung on the front wall.

Kelly and Vic got beers, but Deb wanted lemonade, and the barman found a can. There had been too much drinking, she was watching out for her health. It was already evident that she could roll with things better than the other amigos. She'd seen the meat in the market and turned vegetarian. Not truly, but it was good to have that as an excuse, and Tongans seemed to appreciate hypocrisy.

"What do you give a chicken that lays an egg standing up?" Deb said.

"I don't know, Deb, what do you give a chicken that lays an egg standing up?" said Wichita.

"A standing ovation."

"Oh no, that is very poor."

"My grandmother just sent me that joke."

"A hokey-pokey joke," Vic said.

By eleven-thirty or so, the club was jammed, and a group of more serious volunteers offered a toast to the American Bicentennial. Tonga is where time begins. They were going to be the first in all the world to see 1976, the Bicentennial, one of the volunteers said, at which time a fair-haired New Zealander with mutton-chop sideburns approached the group.

"Excuse me, is it true what I heard," he said, "that your founding fathers wrote the Constitution on hemp?"

"What?"

"No, where did you hear that?"

The New Zealander got a serious expression. "From an American volunteer. He told me Madison and Jefferson were smoking hemp when they wrote the Constitution."

"You heard wrong," and "That is bullshit," the more historically minded volunteers explained to him.

The Kiwi apologized and held up his hands as if he had committed a breach of diplomacy, and kept a straight face as he withdrew, and on this note they set out for the palace.

The Kiwi, Philip English, a teacher at the Anglican school, was called Lampshade Phil because he liked to dance when he got drunk, and he was not the best influence. He led meetings of something called the Royal Nuku'alofa Martini Club, and had urged the amigos to blow off language class. "I believe you'll find that you will need only three phrases to get anywhere here," he said. "Hello, thank you, and thank you very much. *Mālō e lelei, mālō,* and *mālō au pito.* Really, that is more than enough."

Back in Washington, Peace Corps had a hard and fast rule against posting volunteers in expatriate communities. Expats lived apart from their own country, and the host country as well, which meant that they had no code at all, and Peace Corps was nothing if it was not a moral enterprise. Volunteers took vows of poverty and service for two years. They were going to live as the villagers did and speak the language and thereby demonstrate American pioneer virtues to the Third World, even if those virtues had been forgotten back home. Sargent Shriver, the original Peace Corps director and John Kennedy's brother-in-law, had imagined volunteers as missionaries for democracy who would gladly shave their beards and give up sex for two years.

But Peace Corps had had to make its way in the real world by gaining invitations from poor countries, and the King of Tonga needed volunteers as teachers, not as ditchdigging philosophers. He was struggling to create a professional class in the Kingdom, so the Americans, with their bachelors of arts or sciences, could be put to work as teachers, chiefly in the half dozen high schools in and around Nuku'alofa, and Tongan students could then pass the university entrance exam and go overseas to colleges in New Zealand or Australia.

From the Tongan standpoint it was well that most of the 80 Peace Corps volunteers were clustered in the capital. Tonga had maintained its independence through the colonial period by clever statecraft and the advice of a rogue missionary. The only remaining Pacific monarchy, it was deeply protective of its traditions, and as much as the King sought his people's participation in the modern world, he was wary of outside influence. Many Tongans, among them the King's late mother Queen Sālote, had felt that American soldiers had corrupted Tongan morals during World War II. Thousands had been stationed here, and had nicknamed Tongatapu "Fatcat Island" and spread word around the Pacific that the people were friendly, and that the girls' favors could be purchased with cigarettes and gasoline.

The present Queen, Mata'aho, was said to have opposed the invitation to Peace Corps out of concern that Americans would bring permissive ways to the islands.

As they had done. The Americans dressed casually, the men visited Tongan prostitutes, and several volunteers living on an outer island had lately been arrested for growing marijuana. The Crown Prince had rushed with the defense force to the place, and the Peace Corps director, a round-shouldered, culturally sensitive man called Dick Cahoon, had warned volunteers that if there were any more arrests, the entire program was in jeopardy.

The volunteers heard the palace before they saw it. A military band in white uniforms was playing John Philip Sousa out on the great lawn, and a long line of *pālangis* snaked up on to the gingerbread porch. The palace was a Victorian mansion with a bulbous red metal roof. Deb and Vic joined a crowd on the porch and looked through the glass doors at His Majesty's drawing room. A Chinese magician was balancing plates and even a chandelier on canes.

It turned midnight. Fireworks went off and an honor guard of motorcycles rode across the lawn, and behind them His Majesty in a Rolls Royce. Then people started going inside.

The volunteers went by the King and shook his hand, and were floored by the size of it. Like a catcher's mitt, Vic said, and Deb said his hand was like a bunch of bananas, the fingers going up past her wrist. His Majesty was a large and polished man of 57 wearing a white military tunic and a sword, but he had a boyish sense of humor and a boy's energy, too. He was always coming up with ideas to improve the Kingdom. Early on he had reformed the Tongan alphabet to eliminate *b* and *d* (*p* and *t* were good enough), lately he was trying to extend the airport runway so that 707s could land. Some of his ideas were quixotic—exploring for oil in Tonga—but no one questioned his tirelessness.

His people considered the King to be godly. He was the product of two great dynastic lines and was the tallest man in the room; he stood on a little platform to be sure of that. If something passed above the King's head it was *tapu*, it must die. *Tapu* was the only Tongan word in the English language; between his historic voyages in the late 1700s, Captain Cook had brought it back to England as "taboo," though in Tonga, it became *tapu* when His Majesty eliminated *b* from the alphabet. The volunteers had been told how impossible *tapu* was to translate. It meant forbidden, but it also meant consecrated. It was the law preceding all other law on these islands, English law and the Ten Commandments. Turning your back on a chief was *tapu*. Seeing him eat was *tapu*. *Tapu* could have an intangible spiritual quality, with the weight of a curse, but it could also be very mundane, and mean "off-limits." For instance, the pineapples in the field outside the language class were *tapu*; they belonged to someone and must not be touched.

Tapu also governed social relations. It was *tapu* to talk about sex when there were males and females from the same family present, *tapu* for a brother to bare his stomach to his sister. The stomach was considered a sexual part, for it slapped during sex.

When the line was done, the King made a move to go and the band struck up a somber air and everyone stood up except Francis.

"What's going on?" he said.

Finally Wichita felt worldly. "National anthem."

"Ohh—" Francis got to his feet.

But a minute later it was Wichita's turn. The head trainer for their group came up to him with a perturbed look.

"The King does not appreciate what you're wearing," he said.

"What? How do you know?"

"His Majesty said something. It's not a real jacket, it looks militaristic."

Wichita was so mortified he had to walk out. That bush jacket was the only jacket he'd brought, and it was good enough for anything in Kansas. Then again he wondered why he had come all this way, to teach children about phenomena they would never see, like snow, and whole phyla that did not exist, like snakes, when he could be pursuing his own future back in Lawrence.

He was degenerating. Vic had started him smoking cigarettes and Johnson had got him going on boilermakers, and if he rode a bicycle to the wharf opposite the one real hotel a girl would sit on his handlebars and ride off with him. Or maybe not even a girl, but a transvestite.

A week back, a talk on Sex and the Single Volunteer had been offered, for the men anyway, and a group had filed into a back room at the Peace Corps office, and Emile Hons had come in in his white skirt and handmade beads and with a neat dervishing twirl plunked himself down on the floor to tell them about the girls who hung out at Yellow Pier across from the International Dateline Hotel.

"Some are innies and some are outies," Emile said.

The transvestites were called *fakaleitis*. A few of them were quite beautiful and could fool you. Emile advised the volunteers to study their hands. The real girls were called *fokisis*. That was Tongan transliteration for foxes, the name given to the women by the American soldiers during the war. A volunteer only had to give a *fokisi* a little something and she would climb on to his handlebars.

A tall blond volunteer from Sonoma County who was determined not to be the Ugly American had shifted uncomfortably in his chair.

"I've never paid for a piece of ass before," he said. "I don't see where I'm going to start now."

But Emile said it wasn't that bald a transaction. The girls might work for cash when a cruise ship came in, but most of them were more interested in a couple of beers, some respect, a can of corned beef.

The Peace Corps had offered no sex talk for the women, but it wasn't as if they didn't have issues of their own. There were strict limits on a woman's role. On Sunday after the New Year, Judy and Deb wore hibiscus flowers in their hair to church, but a worshiper told them severely they should not be wearing the flower on that side of their head, or maybe not at all, the women weren't sure.

Peace Corps had assigned them all homestays in Tongan *fales*, and after church, Deb went off to Vic's homestay for a picnic, and his Tongan family asked him when they were getting married. That upset her.

"I've been trying to be seen more often with other guys," she wrote to a girl-friend. "That will probably give me the name of *fokisi* (prostitute) because in their culture a girl is so highly chaperoned & then only seen with her husband that any girl seen with a lot of guys can be doing only one thing. It's kind of hard to keep a reputation here."

Judy was taking care of hers. A slender blonde adventurer who'd grown up in a conservative Ohio family, she had ditched her homestay, feeling it was a govern-ment scam, to move in with the Peace Corps doctor. While Deb had dropped her homestay to move back to the guest house. The homestay didn't work. She was expected to make a forty-five-minute commute each morning by bus to language class, even as she prepared biology lessons by the fading light of a flashlight on a mat floor, with six kids staring at her all the time as though she were a television set, murmuring, "Pālangi pālangi."

It was the government's responsibility to provide Deb a house, and a man from Tonga High School brought her around to a little hut near the sea. She walked through it crestfallen. There was no glass in the windows and the shutters had been knocked down by the wind. No electric light, no shelves. She'd been warned that the Tongans sometimes treated the Peace Corps volunteers shabbily because they made so little money, and she went to the office to type a letter to the education department.

"It is my presumption that the Department has no intention of housing anyone in this *fale*, merely from the state of neglect and the deplorable condition this questionable accommodation lies in. There is at least one gaping hole in the bathroom wall as well as numerous smaller ones. The outside appears to have been painted at some time in its existence—"

The government said it would look for another house.

So Deb stayed on at Sela's guest house along with a few other volunteers and bought a black one-speed bicycle for $100. Vic came by and they tried not to spend the night together for reasons of reputation, though sometimes they broke down. He taped a picture of Bogart to her wall and turned on a tape his parents had made by leaning a tape recorder up against the television set and in this way they "watched" Chain Lightning together. They made love and Vic murmured in her ear to keep her voice down.

Afterward he sat in the window smoking a cigarette and doing Bogart lines. "Do you want it the easy way or the hard way?"

Deb said that her dad loved Bogart and she made Vic guess which movie star her father had named her after, but Vic couldn't guess.

She was his girl. They stole away to beaches, and carved their initials in a coconut tree with his knife. She wore her yellow bikini and her T-shirt from the Ram, the beer bar she'd worked at to get through college, and whipped up fruit

salad with her Swiss Army knife. She bragged on his Mafia con
girlfriends back home while Vic said she was the most natural p
gone out with. When the eight weeks of training were done and they headed off
to their swearing-in, she stopped at the mirror in the front room of Sela's and
squeezed a zit. No chick had ever done that right in front of him. But Deb didn't
care.

The swearing-in took place at a garden by the sea. The men wore leather
shoes and long pants, and the big kid from southern California who looked like
Clapton wore a gold plaid Levi's suit and platform shoes. The director, Cahoon, a
psychologist in his big black-framed glasses, was excited for the new volunteers
and the adventure they were setting out on. Swearing them in was his last official
act as director of Peace Corps/Tonga. He was going off to head the Peace Corps
program in Samoa.

They held up their right hands and swore to defend the Constitution against
its enemies, without mental reservation or evasion. "And that I will well and faith-
fully discharge my duties in the Peace Corps by working with the people of Tonga
as partners in friendship and in peace."

Afterward everyone got champagne in real glasses.

But it was a bittersweet day. Two of the amigos hadn't made it.

Kelly had been the first to go. He got some kind of viral infection that the
Tongan doctors couldn't figure out, and carousing with Lampshade Phil, who was
himself leaving after two years, had done nothing to aid Kelly's recovery. Finally
Peace Corps medevac'd him to Johnson Medical Center in Pago and he promised
everyone he would be back soon. But that was the last Tonga saw of him.

As for Wichita, the trainers said that if he was thinking about what he was miss-
ing he should leave before he took the oath. The most effective volunteers wanted to
be right where they were, at the far ends of the earth, and though it crushed Wichita's
spirit to turn his back on a dream that he had nursed from the time that Kennedy
rode by his door, the trainers said that he could always reenter, down the road, and
with that sop, he said goodbye to Deb and dragged his giant duffel bag back to the
airport.

•   •   •

The Friday after the swearing-in, Vic and Deb went to the Tonga Club and Emile entertained them by talking about the time he had danced for the King, but Vic kept noticing Emile's friend Dennis. He didn't like Dennis. Dennis was chunky and blond and hid his eyes behind photogray glasses, carried a knife in a sheath at his side.

"He stares at you," Vic said.

"He's nice," Deb said.

"No, he's not, he's a dirtbag. He never looks me in the eye, and it drives me nuts the way he scratches that ugly fuckin' beard."

Vic could not understand the relationship between Emile and Dennis. They had come out with the same group, Tonga 14, and seemed to go everywhere together. Emile was the life of the party, a Haight-Ashbury shaman, a hipster, while Dennis was introverted. "He's in Emile's shadow, he has no shadow of his own," Vic said.

But Deb liked him. She could see the shy gap-toothed sweetness under his thick beard. Still, he complained endlessly about Tonga, and she didn't understand why he'd put himself in a place that he didn't like, so far away from New York, which he talked about in the fondest terms.

They left the club that night at 2 A.M. and got a cab to Vic's new place in a village called Mu'a, ten miles out of town. The cab was an old Chevy with two Tongan men sitting in the front. The soft fat tires made a ripping sound on the coral road, and Vic slid up against Deb and nuzzled her slanting exotic cheekbone.

"Koe mali?" said a man in the front.

Deb sat up. "No, we're not married."

She started going on to Vic about it. A single girl's life was the worst. He didn't notice but at the feasts, she was always the last to be helped. First any man, then a wife, then her. A lot of volunteers pretended to be married so they could carry on a relationship. Not her. "I'm not going to pretend to be married to anyone—"

Vic broke in on her.

"This isn't the road—"

They had gone right down a bush road toward the airport rather than left toward Mu'a, and that was enough for Vic, his hand went into his backpack and he told Deb that when the car stopped, she should run. He tilted over the front seat and held the tip of his bowie knife against the driver's neck.

"You want her you gotta go through me, now stop the fucking car—"

The car slid on coral stopping, and Deb ran out, but Vic took his time. He held the knife at the man's Adam's apple a few extra seconds and said if he saw him again he would do him like a pig, one ear to the other, then slammed the door and walked backward along the road holding the knife out from his side and making sure the car did not come their way. The man was standing on the brake. The brake lights were lit against the bush. Then the lights blinked off and Vic shouted curses into the night.

He and Deb walked the rest of the way to Mu'a, they got there at dawn, and all the way they argued about Tonga.

Deb had not thought the situation was so threatening, but Vic said Deb was naive. "They call this the Friendly Islands but it's not. It's the same as anywhere else, there are bad eggs here, too, they used to use clubs on each other."

"Those guys were just assholes," she said. "Most people here really like *pālangis*."

"Because you're a good-looking chick, Deb, wake up. Look at you, you're shaking."

It was true, she was afraid. Maybe he was cynical, he said, but she was just a kid from Washington State with her backpack and Swiss Army knife and the government had prepared them for nothing and given them a load of propaganda.

Deb said Vic was too negative. Volunteers who had been here for a while said that everything you put into Peace Corps you got back, you couldn't look to the government to fix your problems. She'd learned that lesson when she was a kid and her parents got divorced. It was all up to you, and the people you met along the way.

"Yeah, well, those guys wanted to hurt you," he said.

"I'm not talking about them. I'm talking about this place. People smile to see you, they say, 'Thank you for working,' and they mean it. It's beautiful here."

"Yeah, I know—it's beautiful. But what is beauty without comfort and a little luxury? It's a rock. On the radio the other day the guy goes, 'Chou En-Lai died,' and that's all. Just 'Chou En-Lai died.' Not where he died, not how he died, not how old he was, not what he did. New York could blow up and we wouldn't even know it here."

"Well, Nuku'alofa could blow up and they wouldn't know it back in New York."

They were arguing about Tonga, but they were arguing about themselves. They were done. She'd written home for Marlboro cigarettes for Vic's twenty-sixth birthday in June, but they weren't going to make June, they weren't even going to make Deb's twenty-third birthday at the end of February. Maybe Vic had fooled everyone, maybe he'd only signed up to get a ticket to an exotic place.

He needed out. He would have to see the new director about that. She was getting here any day.

4

The Tonga Club

One morning late in February, a staggering rumor went out among the volunteers, that the new Peace Corps director, who had arrived a day before, had lately run the Barbizon School of Modeling. That had to be wrong. A woman who'd run a modeling school was taking on the country's largest anti-poverty program? Volunteers stopped in at the office to see what was what.

It was a blustery day. The lagoon had a nasty complexion, the linoleum in the Peace Corps office was streaked with mud from the volunteers' filthy flip-flops. And in this setting it came as a shock to see a beautiful dark-haired woman in her forties who appeared to have stepped off the pages of *Vogue* now moving excitedly around the office. She wore a powder blue dress and a matching blue scarf, her heels were also powder blue. Her features were well-cut, almost too fine, and her dark eyes glittered with nervous energy.

Mary George had grown up in suburban Washington before moving like a wind over the globe: New York, Paris, South Carolina. Now Nuku'alofa. She had been a fashion model on television and a lobbyist on Capitol Hill, and on top of everything else she was a born-again Christian. And yes, briefly, she had run the Barbizon School in Washington.

She was political, that was for sure. She promised the volunteers she would fight Washington for higher wages and the Tongan government for better housing. She was setting up a Volunteer Council so that volunteers could hereafter effect policy, and she told Emile she would give him money for a volunteer newsletter. He had floated that idea to Dick Cahoon, and Dick Decent, as Emile called him, had smiled but nothing ever came of it.

In no time, Vic went around to see her, and was ready to dismiss her as some socialite until she shook his hand and about crushed his bones.

"Yes—Mary—listen—I really need a medevac. I've got bad chills and boils and I'm shaky with exhaustion. You know, Tonga is one of the most dangerous countries in the world—"

Mary told Vic that no one could get medevac'd without going to the hospital first. Vic said that he had been to the hospital, and it felt like a morgue it was so underequipped. They hadn't been able to do anything for him.

"But Vic, I can't medevac you," Mary said. "You have to be evaluated to be medevac'd."

"Who can evaluate me?"

"Only a Peace Corps doctor. He can confirm that you have what you have. We're waiting for another doctor."

The last Peace Corps doctor had just left, afraid of what expatriate life was doing to him.

"There's a Tongan doctor," Vic said.

"It doesn't work that way."

That felt like some political catch-22 to Vic, but he understood that Mary was forceful and it was her job not to lose volunteers. By then Tonga 16 had lost five out of thirty-three and counting, and numbers were an urgent concern in Washington. The cost of delivering a volunteer for a year was about $10,000 (only a fifth of which went to the volunteer), and any time a volunteer quit and flew home, it drove that cost up.

Judy left to teach at a high school in the northern island of Vava'u, and the government found Deb a house in the bush at the east end of the city.

Her village was called Ngele'ia, pronounced with a soft Polynesian *ng* sound. Two little white tin-roofed boxes sat together at the far end of a rugby field on the grounds of a primary school. Their front doors looked out on a narrow dirt track, and just past that, mango trees, hibiscus, manioc, taro, and a *faletonga* or two with kitchen fires going in the morning and night.

Emile Hons had the smaller house to the south. He'd lived there for a year and called it the suburbs. He'd fixed up a copper coil contraption so that he could take hot showers, and by placing his radio in a box wrapped with copper wire with the other end of the wire in a coconut tree, he could pull Top 40 from Guam as he ate his cheese omelet supper or hoed his garden of peanuts and tomatoes.

> *Fly robin fly*
> *Fly robin fly*
> *Fly robin fly*
> *Up up to the sky!*

Deb decorated her one room in Tongan style. She laid woven mats on the cement floor and hung the walls with tapa, the cloth made from the beaten bark of mulberry trees and decorated in sepia inks. She got running water from a spigot by the toilet in the corner, and she did her wash in a big basin. Because the harsh detergent did a number on her hands, she put on a Beatles tape, or Jethro Tull, and danced barefoot on the clothes for 15 minutes.

Though usually when Emile came by, she was listening to classical music and reading on her bed.

"Look, I got a cat to get rid of the rats," she said.

"More like a kitten," Emile said.

"Yeah, the rats are going to laugh hysterically if she tries anything."

Emile showed her how to put the feet of her bed in tuna fish cans filled with oil so that insects couldn't crawl up when she was sleeping, and her mother sent her old sheets for the bed, and also her *Schaum's Solved Problems* books, for parasitology and chemistry, and her red paperback BBL manual of microbiology. A dozen Bic pens and books of recipes, too.

In addition to upper-form biology, Deb was teaching Tongan girls how to cook. Tonga High School had a new outbuilding with sewing machines and modern stoves. The big yellow school stood across from the Royal Tombs at the edge of the old city and was the leading government school. His Majesty sometimes said that his greatest achievement was starting the school in 1947, when he was the Crown Prince under his mother, Queen Sālote, for it had given so many top students an opportunity to study overseas.

Deb had to admit that teaching wasn't her forte. Everything had to be written out on the blackboard for rote study, and when she refused to do that, she could see where some students got upset with her, or confused. She said she was trying to teach them to think independently, and then they reacted like she'd said something sacrilegious.

The girls' conditioning was pathetic, she told a girlfriend back home. "They don't even want to have any initiative or independence of their own," she said, and though she tried to get them to think about having control over their lives instead of being slave labor, all they seemed to want to do was get married and have babies.

Emile taught art at Tonga High. At heart, he was a California bohemian. He moved easily in the Tongan demimonde. The *fakaleitis* who hung out across from the International Dateline Hotel did a cross-dressing fashion show for him, and he took their pictures. From time to time a *fokisi* came to his *fale* to spend the night. Deb teased him about it, and if she thought he wasn't telling her the truth, she rubbed her nose with the tip of her finger.

Emile wrote home about her. "I've got a new next-door neighbor. Nice, and she's teaching at Tonga High and has huge . . . eyes.

"That doesn't sound right. Huge eyes, great smile—and a boyfriend! One can only hope!"

Deb dressed modestly, in denim skirts and men's buttoned-down shirts, still men noticed her big laugh and the way her body moved. A number of them had active crushes on her, and some of the women said Deb needed attention. They probably misread her. She was outgoing but she had a solitary nature. She had a wide circle but didn't go in for intense friendships. In college she'd been happy sitting in her room on a Saturday night, reading a novel or studying.

Still, now that she had split up with Vic, she felt social pressure. "It's too hard to get serious here because the ratio of girls to guys is small & we have to be shared," she wrote to a college girlfriend, half-joking.

* * *

Just across a lot from the Peace Corps office on the shore of the lagoon stood a rambling wooden building called the mosquito lab, where older volunteers gathered. Tonga 11 had come out two years before Deb's group, and a number of them had lately extended their service. The three who lived at the mosquito lab, Jackson, Wyler, and Bevacqua, were considered to be *fakatonga*. They liked Tongan ways.

Frank Bevacqua was a moody and artistic New Yorker with curling dark hair that fell over a strong forehead and prominent nose. He had signed up for the Peace Corps during the Watergate hearings in 1973 when he said to his then-wife, "We've got to get out of this place." After two years in Tonga his wife had gone home, but by then the world had peeled open to Frank the way the founders of Peace Corps had promised, and he couldn't understand why anyone would want to leave. It was lovely and strange out here.

Like the other Tonga 11s, Frank had spent his first two years teaching in outer villages before moving to the city. Now he was working for the government newspaper, the *Chronicle*. He liked to sit out on the steps of the mosquito lab with a guitar, and one day in March Deb came out of the Peace Corps office and started talking to him about the Peace Corps administration and how much money the Peace Corps director made living in her fancy house.

Frank asked if she wanted to go for a swim later. She said yes.

Deb liked Frank at once. He was sincere and emotional, and though he could be irritable, he could turn giddy in a second. Deb's laugh did something to him, its strength, its femininity. He liked the fact that she didn't take things too seriously but teased him and got on top of him and tickled him till he curled up in a ball.

She wasn't caught up in her looks either. Once he was playing with her hair and said that everyone said she was beautiful. She looked away, embarrassed. "I know some people say that, I hope they think more of me than that."

Deb started hanging out at the mosquito lab. Jackson would make banana pancakes for breakfast, and at night Wyler showed up with *hopi*, mango home brew, and they played poker, and the men were more comfortable cursing one another in Tongan than English (*Kai ta'e, Tula!*—Eat shit, baldie!) or throwing masturbation jokes at one another.

"Oh yeah I saw your girlfriend today, *Mele Nima*." Mary Hand.

If it was Sunday, a couple of the men put on *ta'ovala*, the traditional woven mats Tongans wore over skirts, and went to church. The mats hung from pegs above a pile of diving equipment, flippers, knives, masks, spearguns.

Frank was a lot older than Deb, 29 to her 23, but he soon understood that she'd had a tough growing-up. Her dad had hurt her. He saw her little blue notebook with its addresses. Everyone in her family was right in the front. Her brother, her mom, her aunt and uncle. Grandma and Grandpa Gardner in Washington. Grandma and Grandpa Johnson in Wisconsin. Not her father. He had a place all to himself inside the back cover. Dad. SRA Box 25A. Anchorage.

At the school break in April, Deb took a boat to Vava'u to visit Judy and they talked about how far away they felt from home and how the distance was necessary. "I hope you understand that as far as my growing up and mental health goes, I had to go out on my own," Judy had written her parents. "I'm trying my hardest."

The two tomboys took an outrigger dugout canoe into the harbor and worked on their J stroke together, and Deb saw the romance of life on an outer island. Judy had eaten fresh whale meat and gotten lice, and her school wasn't nearly as demanding as Tonga High. In the library the books were shelved according to the color of the spine. One shelf for red books, another for blue.

When Deb got back to Nuku'alofa, the Tonga 11s encouraged her to get out of the city. That was the real Peace Corps experience, living in a village with maybe one other volunteer, the rest of your neighbors Tongan. There were long days of nothing to do, and everything you accomplished was up to you to figure out. Deb could handle an unstructured life. She liked to sit in her *fale* and read all day, she could make do on a little island.

The third-year volunteers pushed Deb to avoid the American gang in Tonga 16, the volunteers who thought a lot about the United States and didn't really want to assimilate here. When a plane went overhead, Francis Lundy could tell you what flight number it was, and how long it would take to get to Honolulu or Apia or Suva, Fiji, and how long before you were back in the States. Francis was dating Vickie Redpath, a short, blunt-speaking volunteer from Pennsylvania who had worn a stylish short dress to the swearing-in when just about every other woman had on long dresses, and together they put down *pālangis* who were trying to be *fakatonga*. Dave and Kay Johnson never seemed to have left southern California and though Paul and Laura Boucher were committed to international service, they kept a sophisticated household of the sort they had probably had back in Connecticut. They both taught English. Laura worked at the Anglican school, and Paul sat in the Tonga Club with a very thoughtful look on his face, writing in his journal, or talked with Vic about Chick Corea and Weather Report.

Deb had friends in the *fakatonga* gang and among the America-oriented too. Frank thought she was struggling to grow up. She was trying lots of different ways, and maybe other people were going to judge her, but she didn't care what other people thought.

He stole hours with her in the afternoon at her hut in Ngele'ia, with the musical sound of women banging tapa hammers in the bush, and one day Deb told him the story of her cold war with her father.

It had started over little stuff. Her brother Craig was the year after her at Washington State. Her dad and his new wife Barbara were paying for the kids' college, and they forgot to mail Craig a check, or, according to her father, the check got lost in the mail.

It wasn't the first time her father had seemed insensitive to their needs, and Deb took up arms for Craig. He was eighteen months younger and they were close. When their parents split up, they'd vowed never to split themselves. They'd gone backpacking and bicycling together in Europe two summers in a row, and when the check didn't come, Deb wrote her father. Probably she should have called, but she wrote.

"Dad, you are treating Craig like shit."

It was disrespectful, but she'd always said anything she wanted to her father. Her father was a big impulsive man, and he wrote back to Deb in even sharper terms than hers. He'd heard that Deb was working at that beer bar, the Ram.

"If I'd known that was what you were planning with your life I should have just sent you to bartending school," he said. "You want to be on your own, you're on your own—" And he cut off her funds.

She felt so deeply wounded by that. A year later she joined the Peace Corps and didn't even tell her dad. "We're having a cold war, he can find out about Peace Corps from Craig."

Still, the war was on her mind.

Her Grandma Agnes had just found out about the fight from Craig at Christmas, and she wrote Deb a heartrending letter. "I wish there was a better understanding between you two and your Dad, Debbie. Some day this will be remembered for you all, the saddest Christmas ever. . . .

"I think about you so far away, in a new land. Any part of God's country is of some interest to me. I wish I could be a bird of knowledge, I could fly, bring back some knowledge of every nation and country."

Grandma Agnes was married to a hardheaded Gardner herself. She was telling Deb not to be so stubborn, but to open her heart and let her spirit go to Alaska, make up with her stubborn dad.

It was probably inevitable that another of the Tonga 11s who came around to the mosquito lab would develop feelings for Deb, and Deb for him. The problem was that he was one of Frank's closest friends, Mike Braisted.

One Sunday night at the beginning of May there was an impromptu party at the mosquito lab with several bottles of cold Steinlager, and Deb and Doug Jackson got into monkeyshines together. She teased him about swimming naked in the lagoon.

"It's *tapu* to go skinny-dipping in Tonga, Deb, don't you know that?" Jackson said with a straight face. You could only go skinny-dipping on a remote beach, or if you were out on a boat.

Deb gave it right back to him. He went swimming naked off his boat, he should go right now. It was dark, she'd go with him. She stood up.

"Come on, Doug, let's go, let's go swimming now, get out of those stinking shorts—"

Till Jackson broke into a mellow grin and said, "That's OK," and didn't take the dare.

But Mike Braisted went swimming with her, and for a couple of hours they tried to drown one another in the lagoon.

Mike's hair was bleached blond from all the time he spent in the water. Tall, skinny, reserved, and masculine, Mike had grown up in Venezuela, then Colorado. He liked Tonga. He felt that the Tongans had taught him important lessons about enjoying the moment and not worrying so much about getting places.

Mike was 25 and very reflective. "I don't like males, volunteers and Tongans, talking about females as inanimate objects such as holes," he wrote in his journal. "I know I would dislike to be considered as a being assembled just for the several inches of gristle and what it can do. To me a woman is an individual in thought, action and choice. And that might explain why I'm slow on the move. I don't think I should have to do it all."

He didn't with Deb. She poked him in the ribs and made fun of him, and Mike felt completely comfortable around her. She never wore makeup—he didn't care for makeup—and he marveled at how much feeling she could get across with her hands and eyes. When she talked about her brother, Craig, and how close she was to him, Mike was moved by the depth of emotion.

Mike worked for the Wesleyan school system, and tried to help Deb through her teaching troubles. He told her about all his problems teaching science his first year. Good friends of his had dropped out, and he became discouraged. He didn't like cooking for himself, he was lonely and sometimes bored. The language was frustrating. But in the long run, the people, the ocean, and the deep satisfaction of self-discovery had pulled him through. Deb had the resilience to make it, he told her. He might be better at the language, she was better at cooking.

Mike had begun seeing Jen, an Australian volunteer, but it seemed like something might happen between him and Deb, and much as Mike disliked jealousy in others, he began to feel jealous toward his friend. Frank and Deb were close but they seemed to be able to come and go without either one feeling put upon. And Frank was going back to the States soon, for a one-month home leave.

"That which I'm afraid of has happened," Mike wrote in his journal. "There are two people that I care an awful lot for and I cannot make a decision. I tried to explain this to Frank but I failed miserably. I sounded very sure of myself and this sounded very conceited, something I'm not. I think I sounded like I thought I could wrap Debbie around my finger while I was being wrapped around her little finger, which to Frank must have sounded extremely strange especially since she told him just last night that she is in love with him. I know that, yet still feel that perhaps she and I can become closer?

"Maybe I want to feel that way because otherwise it would mean that she is unconsciously using me in an attempt to forget Frank. Since he is due to leave shortly. That would hurt—"

May was Tongan fall. The smell of ripening guavas filled the air, and Frank Bevacqua and Greg Brombach, the big volunteer who looked like Clapton, sat in the Tonga Club with guitars, trying to remember the words to Hendrix's bad-time ballad, "Hey Joe."

> Hey, Joe, I heard you shot your woman down,
> You shot your woman down . . .

Other volunteers called out phrases and between one and the other, they came up with almost all the song.

> Hey, Joe, where you going to run to now—
> Way down south, way down to Mexico.

Then someone came into the Club and said a truck had pulled into the Peace Corps lot with a red trunk on the back. The Johnsons' congressman had delivered. Dave Johnson unscrewed the lid and lifted out his state-of-the-art stereo equipment like worship objects, and Mary George declared that they would have a party at the Peace Corps offices.

Johnson built seven-foot wooden cabinets for his JBL speakers, and he and Brombach mixed a killer party tape, with peaks and plateaus and crescendos, to get the dancers higher and higher, and they had a blowout bash on a Friday night at the end of the month. The song of the hour was the reggae anthem "I Shot the Sheriff." The volunteers heard it several times that night, Clapton's version, and Marley's, as well as the whole Sticky Fingers album, with Mick Jagger doing a rockabilly love song.

Take me down, little Suzie, take me down
I know you think you're the Queen of the underground. . . .
And I won't forget to put roses on your grave . . .

Deb and Mike Braisted danced ten dances till they were so sweaty they went into the lagoon and swam way out to get past the reach of Elton John singing "Hercules," and later when they came back, their hair was wet against their necks but their clothes were only blotted with water, and people said they'd jumped into the lagoon naked. There was gossip. Maybe someone had even seen their bodies flashing in the moonlight.

Deb knit Mike a wool cap. He took a boat to Vava'u for his job and lay on the deck with the hat on and wrote her a love letter.

"Hi there, Sparkle eyes, I guess you don't know about that, do you? Well, when you look at someone, someone I'm going to assume that you are happy to see, your eyes take on a lustre of a sort that kind of seems like the essence of life is emanating out.

"Whew! Got a bit heavy there, but then warm thoughts generally come out of a warm head."

Then he scratched out a self-conscious postscript: "I hope this doesn't cause any new hassle between you and Frank."

Mike was in turmoil. He could see that Frank was confused, as well as his girlfriend, Jen. Mike caught Jen looking at Deb critically, and saw Deb getting flustered by sharp things Frank said to her. Then Mike wondered if he should show his journal to Frank, to let his old friend know what was in his heart.

In late June, the blossoms covered the mango trees like a kind of rust, and Mary gave Emile a little money to put a Bicentennial newsletter together on the mimeograph at the office. Emile drew a caricature of George Washington on the cover, but wasn't sure of the headline till Dennis came over to his house and saw the picture on the table.

"Washington. First in War, First in Peace, Last in the American League," Dennis said. "That's what we used to say in New York when the Senators were still in Washington."

"Great," Emile said, and wrote it out.

Emile and Dennis were not as close as they had been their first year in Tonga. Dennis still came to visit Emile in Ngele'ia, and still had a key to Emile's place, but Emile wasn't sure whether Dennis was coming to see him or Deb.

In his photogray glasses, with his knife at his side, Dennis was socially clumsy but generous. If someone was nice to him he would do anything in return. Deb rode out to the beach with him one Sunday, and later he installed a sink in her kitchen counter. She had to carry the water from the spigot in a pail, and the waste line just dumped out into the rugby field, but it worked. Afterward, Dennis felt that Deb was not thankful enough. He seemed to expect some intimacy to result, but it didn't. She liked him, but she wasn't attracted to him. She told others that he was oafish.

Deb and Dennis shared an irreverent sense of humor, but they would probably never achieve rapport. Dennis was a brooding New York Jew with a rich interior life, his mind moved freely through the scales and caverns of mathematics. While Deb, raised Lutheran at the other end of the country, had a richer exterior life. She was taken up with the role of women in Tonga and the shortages in the stores. Dennis helped her grade her sixth-form biology papers on a 60-point system, but he said Deb was insincere. Under the gruff crust, he could be extremely sensitive. He was afraid to look her in the eye. Still he thought he should get a turn with her.

Dennis taught math and science at Tupou High, a Methodist school and the best school in the land, but he was cynical about Tonga. The rituals and moralism seemed primitive to him. He bragged that when students didn't understand something he told them to jump in a lake. That offended the tall blond guy from Sonoma County—it was Ugly American, there were no lakes in Tonga.

But Dennis's sarcasm and intelligence made him popular with the "faka'Amelika" volunteers. Paul Boucher was the best basketball player on the island and Dennis talked to him all the time about sports, New York stuff, things they missed from the East Coast. Paul taught English alongside Dennis at Tupou High and thought of him as funny, calm, together. Francis Lundy had liked Dennis from the moment that he got a letter from a friend in the small town of

Detroit, Texas, and Dennis had leaned over his shoulder in the Peace Corps office and flicked the envelope.

"You Texans, you think you own everything—"

July was Tongan winter. The volunteers slept under wool blankets, and Dennis awkwardly invited Deb to come over to his house for dinner a week after the Bicentennial, and she accepted. As Dennis pined for her, Emile encouraged him to try, and the American gang helped him put the meal together. Emile thought of it as a high school gambit, and the others also saw the date in high school terms: Deb Gardner really ought to tumble for Dennis Priven. It turned into a big deal. The Bouchers helped Dennis cook the food, and Vickie and Francis loaned him some fancy wine glasses they had. The Johnsons helped too. Perhaps implicit in the planning was a judgment of Deb. Dennis was a serious soul, Deb was a party person. He'd be good for her. Maybe there was some cultural misunderstanding, too. Paul Boucher and Vickie Redpath were East Coast intellectuals. So was Dennis. They weren't as familiar with Deb's type, a girl from the Pacific Northwest with her Kelty pack who had grown up camping with her father and was smart but not overly analytical.

Francis expressed misgivings. Deb wasn't interested in Dennis that way. The party was a dishonest setup. Should people who knew and liked Deb be helping Dennis this way?

So it went. The dinner came off badly. Dennis had high expectations and had gotten Deb some sort of gift, spending real money. He was full of awkward feeling and the situation became unpleasant. She ran out of his house, got on her bicycle, rode into the night.

When Deb saw Frank later, she was upset. "He must have spent $100 on this dinner and doesn't he know I don't want to go out with him?"

"You have to tell him that."

Frank was hardly disinterested. Deb was moving on. She'd told Frank she felt overwhelmed by their relationship. "This is getting too serious, I need to be free—"

He had a hard time hearing that. She was a strong woman, and he built her up even more in his mind. No, she didn't hold her beauty over men, but it wasn't as if her Cherokee looks and spirit didn't grant her a power that she enjoyed.

Everyone was wondering how many guys would try to make their move when he went home on leave in August.

Before Frank left, a Belgian doctor who had come out to work in a government clinic threw a party at his house by the sea, and Deb danced with Mike Braisted and Frank sat there with his heart hurting. The party went in and out of the house and onto the beach. The doctor served a wicked fruit punch, a few cups made the top of your head come off, and he played James Brown and Santana on a reel-to-reel tape deck. Frank sat on a couch stunned by the punch and the sight of Deb twirling in and out of the dark to long spells of Carlos Santana's guitar. Her mouth was open in laughter, her cheeks were dark with color, and thick strings of hair slapped around her face, slicked by saliva, sweat, life juice.

Deb Gardner was so alive, she seemed to be growing out of Frank right before his eyes, and that hurt more than anything, that he wouldn't be around to see what came next, who she really was.

She asked him to do her a favor on his trip home. Frank was going through Seattle. She wanted him to visit her mother and get some stuff she needed, bring it back with him in September. Recipes, spices, textbooks. "Also any recent copies of *Cosmo* or *Vogue*. Forty females would love you," she told her mom. "Be super nice to Frank, he's a special friend …"

She wrote out her mother's phone number and address for him, 2822 Garden Court. Steilacoom.

5

Steilacoom

Steilacoom was still as a painting. The azaleas were ablaze, the barkless madrona trees gleamed in the sun, a wooden sign said it was the oldest town in Washington. As I wound through town, the lush foliage parted now and then to afford a view of Puget Sound, its waters glinting narrow, deep, dark blue.

Garden Court was a tidy subdivision, a dozen beige structures in a boxy troop going down a wide curl of asphalt.

Alice Gardner was long gone from 2822. All four units looked empty, a FOR SALE sign hung off the carport. One unit had collection notices for a man called Trevor Lyons stuffed inside the storm door and every available crack.

Alice had probably remarried. At the Seattle Public Library a woman at the genealogy desk wearing a bunch of silver New Age jewelry told me that divorcees in their forties almost always did. She reached for a book called *Finding Anyone in the 20th Century*, and flipped through it, past photographs of a driver's license and obscure public records.

My spirit quailed. Finding Deb's mom was going to be a long, bureaucratic grind.

But the next morning after only five minutes at a microfilm reader in a warehouse-like public building, "Gardner, Alice Marie" appeared in script in a column of Ladies' Names in old marriage records. Four years after her daughter's death, Alice had married a man in the logging business.

I grabbed a Tacoma phone book. Alice and her husband lived on the north side. A biggish intersection. A blue ranch house. A half-dozen rhododendrons. A pretty gray-haired woman in a blue velour tracksuit opened the door but held the storm open on the chain. She looked to be about 65. Her hands were fine, like her daughter's.

"I'm researching a book about the Peace Corps—"

Alice stood back as though I had pushed her, then opened the door for me, let me in. I came into a room carpeted in powder blue with couches along two walls, and Deb's senior-year portrait from Washington State on a sideboard.

"John," Alice said, going to the head of the hallway. "John, please come in."

A lean and still athletic man who looked about 75 came into the living room. Alice introduced me.

"May I see your ID?" Alice said.

I gave her my New York driver's license, she passed it to her husband before handing it back.

"I just wanted to make sure you're not Dennis."

Alice and I each took a couch, John the recliner. She held a handkerchief to her eyes and said that she had never learned what had happened in the case. It had been the worst year of her life. The Peace Corps had said that Dennis was going into a mental hospital, she hadn't heard a word after that. She had trusted her government, hadn't pressed for answers.

Suddenly I felt the need to be very careful what I said, and I got out a lawyer's memorandum from my old file. It related an encounter at Peace Corps headquarters on Dennis's return from Tonga.

"Priven came to me and stated his intention to return to New York immediately. He demanded his plane ticket. . . . He stated that he did not wish to be placed in a locked ward because of his recent incarceration, he needed some freedom. Since we have little choice in this matter it seemed that the best course of action would be to allow Priven to stay at the National Hotel on Friday night while we kept some sort of informal contact with him."

Alice took a deep breath.

"So—he spent the night in a hotel."

"He was back in New York in a couple of weeks."

She covered her eyes for a minute but when she took the hanky away she seemed to look at me for the first time, calmly.

"I wonder what else he might have done," she said. "I think about the other families . . . Did you know that Debbie was afraid of him, and asked for a transfer?"

"No. Can I write that down?"

She held up her hand, said she wasn't ready to talk to a reporter. But I sat forward on the couch.

"Alice, what happened here was wrong. I want your help. I don't want to cause you any more pain, but I must tell you, I am going to write a book whether you help me or not. It's a public matter and no one knows about it."

From the recliner, her husband took over.

"Of course we've always wondered what happened. But things like this go on all the time. Look at O. J., or Ted Kennedy. What is the value in your doing this? Nothing will be done."

I said the case haunted people. Old Peace Corps volunteers had never been told what had happened. They still wanted answers, and they deserved them.

"But it's been twenty-five years," he went on. "Debbie's name wouldn't even ring a bell. No one thinks about this anymore."

"I think about it all the time," his wife said.

She asked if I knew anything about Deb.

"She was idealistic, and she was beautiful."

"She was very pretty," Alice said, as if still called upon to check her daughter's vanity.

But she stopped herself from saying more. Craig would make the decision about helping me, Craig worked as a wildlife biologist in Alaska. Alice wrote out his name and address in fine clear script.

"What about Wayne Gardner, is he alive?"

"Wayne—I'm sure he is, I'd have heard if he'd died."

Deb's father lived somewhere in Idaho, Craig could tell me where. Craig had recently gone hunting with his father.

Craig worked for the Alaska Fish and Game Department in eastern Alaska. His picture and some of his writings were online. He was a lean, bearded environmentalist, and extremely independent. He had spent weeks at a time tracking wolves and caribou in the Alaska Range and made a study of the solitary wolverine, which he described as the most beautiful creature in nature.

I wrote him several times, and heard nothing. I wrote back to Alice and heard nothing from her. I sent Craig a letter for his father and included a stamped envelope for him to forward it.

A year passed—more. I didn't like to think what would happen if they asked me to stop, but they didn't, and meantime I met more and more people who were willing to help even when I told them that Deb's family wasn't.

The men who'd loved Deb Gardner were angry at first. Vic Casale grilled me for an hour on the phone from Japan about my motivation. "Who are you? Why are you doing this? What are you doing in my life?"

Frank Bevacqua cried out on the phone. "I'm shocked. I'm very concerned about this. I'm overwhelmed at the moment. You're going to have to excuse my paranoia, and skepticism. I'd like to know—I'm sorry to be melodramatic—what do you have so far and where are you going with it?"

Emile Hons had stopped returning my calls. Then I announced to his answering machine that I was going to be driving through California in August 2000, and he called back and said he could give me thirty minutes at his office on Saturday. He sighed. "I don't know if this is the smart thing, but it's the right thing."

We talked for five hours, and it was the first of many visits.

Once at Emile's place in Marin, I met Mike Braisted. He was broad-shouldered, quiet, and when I said something about people's characterizations of Deb's sexuality, his face flushed and he stood up. For a second I thought he was going to hit me, but he just walked out of the room.

Another time, there was barbecue and slides. Emile had asked the seven volunteers who were coming if they minded if I came and they said no, but while two were friendly they wanted nothing to do with my work, and a third was on the fence. Emile said he felt they were worried about how much this book would hurt people, hurt the Peace Corps. Many former volunteers look back on their service as the finest chapter of their lives. Here I was, pawing away at one sensational part of that experience.

"Peace Corps is the number one or number two best thing that America has done in its history, and something like this might tarnish it," said Alan Burrus, a former Tonga volunteer who was on the board of the National Peace Corps Association. "This is a very personal situation," said Laura Koutsky, a member of Tonga 16, now a professor of epidemiology. "It's not something that anyone else should be concerned about, it's just like if my husband was walking down the street and was hit by a car." She asked me how I slept at night and called me a muckraker, a sentiment echoed succinctly by another volunteer who declined my overture. "You're stirring up a mud puddle that settled a long time ago," she said.

Some of these people were afraid to help me. "You found me, so can he," said one. "I have young children," said another.

But many more former volunteers did help, and Bob Forbes explained why. "This project is providential—it answers the need felt by a large number of people that this be opened up at last, and closed."

Vic Casale and Frank Bevacqua had never lost their photographs of Deb or their writings about her, and now they shared them with me. Months after he'd clammed up in Marin, Mike Braisted surprised me by pushing his journal to me across his dining room table. Once when I came to the mall he managed, Emile told me about a dream he'd had the night before.

"I was walking on the sand dunes early in the morning near Santa Cruz, a place I used to dig up fossils when I was a kid. Three people were coming along the dunes, Debbie and her boyfriend and a friend of the boyfriend. I felt such a need in going to talk to them, to warn Debbie. Even though it's now and Debbie died so long ago, in the dream I felt that I could save her. So I went up to the boyfriend and said, 'Excuse me, there's something very important that I have to say to her, it's very important,' but the boyfriend told me to get lost in no uncertain terms, and yet I still felt such urgency and I pulled the friend aside, and I said, 'Please, there's something very important that I have to tell that woman that I know.' The friend agreed to talk to the boyfriend and try and get him to change his mind. He walked away. I sat down on a box of beer."

It worked. Deb Gardner came walking back across the dunes, Emile rose to warn her. And then he woke up.

6

The Perfect Volunteer

John Kennedy signed an executive order creating Peace Corps on March 1, 1961, and the program was overwhelmed with applications. It was at the forefront of the culture, and ambitious kids wanted to join, Ivy League kids, kids who saw their future in academia or politics or business and understood Peace Corps to be a gold star on the résumé.

The founders of Peace Corps believed that they could change the world forever by sending forth an army of 100,000 kids who hated colonialism and would live in chicken coops, but from the start Congress expressed concerns. "Even if these people were selected by an electronic brain of absolute perfection, there will be some misfits," Hubert Humphrey said. Sargent Shriver assured Capitol Hill that the volunteers were spiritual people of sacrifice and maturity, while secretly the Peace Corps worried about the issue Humphrey had raised.

The first answer was boot camp and organizational science. The Peace Corps leaders possessed that 1960s hubris that psychology could do anything, even find the perfect volunteer. Training went on for nearly three months on American campuses that were teeming with psychologists bearing clipboards. The would-be volunteers were observed, interviewed, graded on a 9-point system—7s were fair, 6s were weak, but 5s were the inspired madmen, 5s were high risk, high yield—and given the Minnesota Multiphasic, as well as other, more practical tests. They had to kill a chicken, had to sleep in abandoned buildings.

Or a man drifted up to the other cots, those of the effeminate men anyway, and said, "Hey, do you want to come to a party?"

The trainees who said yes were sent home, gone before the others woke up, so that their departure would not hurt morale. The Peace Corps word for this was "deselected." Deselected for declining to eat cold leftovers. Deselected for not accepting a rabies shot to the stomach. Deselected for choosing a relative's wedding over a day of training.

A professor gave a lecture on the domino theory and a kid in the back cried, "This is bullshit!" and was deselected.

And always gone before breakfast, so as not to sow dissension. But what did that nicety matter? Deselection damaged morale. Volunteers never forgot the great people they never saw again for some arbitrary reason or another—that kid with the agricultural degree and bright blue eyes who was thought to have too much missionary zeal—and they held contests to see who would spend the least time with the psychiatrist. The psychs also saw the weaknesses in their method. They knew they had no real tools for predicting who would flourish and who would founder in a Third World village.

The best test was peer review. "This is a necessary evil," the shrink in a cowboy hat and cowboy boots and big cowboy belt announced in a booming voice. "We need you to write down, who are the three people who will do best overseas, and who are the three who will do worst, and why."

But when the volunteers saw what was going on, they refused to rat one another out.

Then the country changed. The dream of 100,000 volunteers evaporated. By 1966 Peace Corps reached 15,000 volunteers, but after that the numbers started falling, and when Nixon came into the White House, they slipped further. Nixon saw Peace Corps as a haven for Democratic kids who opposed the war and wanted to burnish Kennedy's image, and Peace Corps advocates began to fear that its enemies had the upper hand, that the program was being left to die in the straw in an out-of-the-way place.

Nixon ended Peace Corps's independence by burying the agency in a volunteer bureaucracy called Action, and his successor, Gerald Ford, who had never voted for

Peace Corps when he was in the House, began cutting back the program's $81 million budget. Peace Corps could no longer afford all those fancy psychologists, and the new director, a religious man in a bow tie, assured Congress it didn't need them anyway. Psychologists had never been able to tell who would make it living in a mud hut.

By then, a great number of former volunteers had come back to staff the Peace Corps headquarters across from the White House, and they knew what made a good volunteer. Volunteers weren't ambitious people who were thinking about their careers back in the United States. They didn't really care for mainstream values. They were adventurous, they were informal. They knew how to live in the moment. It was the reason westerners tended to do so well, people like Doug Jackson and Mike Braisted, the Tonga 11s who couldn't care less that only 200,000 people in the world spoke Tongan.

The first volunteers had been conformists. The new volunteers were 1970s kids, a little alienated.

Nixon had just resigned in digrace when a new group of teachers called Tonga 14 gathered in Denver in October 1974. There was a psychologist in the hotel; each trainee went to see him in his room.

"Would you like a drink?"

Emile summoned his wits. It was ten in the goddamn morning. If he said yes the shrink would think he was an alcoholic; if he said no he would think he was antisocial.

"No thanks, I had a few last night, we went out to the Playboy Club."

"Yes, I heard—you're quite the ringleader."

Well, Emile explained, he had a Playboy Club key. And that was it from the psych, ten minutes and on your way. By that point volunteer numbers had dropped to 6,000, an all-time low.

On their way out Tonga 14 stopped in Fiji and got a bus across the main Fijian island to the capital, Suva.

The road passed through mountainous villages, and the Bird Lady looked around at the other volunteers and saw kindred spirits, people like her, smart, unconventional people, maybe a little odd too, people who had their own good reasons to go to the ends of the earth. Emile was the life of the party but his humor sometimes had an edge, as if to guard his privacy. Tom carried *Be Here Now* by Ram Dass in his bag, and when the psych had asked how he would handle two years of celibacy in a Christian country, he'd said, "No problem, I'm celibate already."

Dennis was shy, mathematical, muscular, with deep-set eyes, a bunch of rubber bands on one wrist, and a high voice with a Brooklyn accent.

And Barbara Williams was known as the Bird Lady because she was so focused on birds. Her first memory was of evening grosbeaks at her grandmother's house; she'd been netting birds at Cape Blanco when she found out that Peace Corps had accepted her. The thick glasses she'd worn since she was a girl had made her feel socially awkward, even at times incompetent. She curtained herself off with long thick hair, and studied the mynah birds and epiphytic ferns outside the bus window. Still, Emile and Dennis noticed her slender body and delicate coloring, and murmured about how she'd look with those glasses off and the hair away from her face.

Suva was a real city. The Bird Lady found a red-vented bulbul's nest, Tom went to an ashram, Emile and Dennis went down to the Indian shops. Emile looked at camera equipment he couldn't afford, and Dennis staggered him by figuring the exchange rate instantly before purchasing a fancy Australian dive knife with a six-inch-long serrated blade and a black handle and a shiny steel bulb at the handle's base for banging the airtank to get other divers' attention. A Seahorse, it was called.

They had bonded. Emile was smart enough to be Dennis's friend, Dennis was funny enough to be Emile's. He was just very shy, and that voice—was it a lisp, or was he tongue-tied? Emile wondered if something inside his friend held the reins of his tongue.

The idea of New York scared Emile, but Dennis told him how wrong he was. He walked anywhere he wanted at night in Brooklyn, and yes, the Jewish kids fought with gangs of Irish kids and Italian kids from different sections of Sheepshead Bay, but it wasn't lethal. The next day, on the flight into Tonga, Dennis did the accents for Emile, the old Jews, the young Italians.

"This Italian scientist was in a lab, correlating the number of legs a frog had to how far it could jump—"

"All right," Emile said, smiling already.

"He puts the frog down on a table, pounds the table. 'Jumpa froggie jumpa!' The frog jumps fifteen feet. The scientist says to his assistant, 'Laba assistanta, put in da manual, da froggie with afoura legs he-a jumpa fifteen.'

"He chops off one of the frog's legs, pounds on the table. 'Jumpa froggie jumpa. Laba assistanta put in da notebook, da froggie witha three legs he ajumpa ten feet.'"

Then five feet, and two feet, till the scientist cut off the last leg.

"'Jumpa froggie jumpa!' Hits the table again. 'Jumpa froggie!' Nothing. 'Laba assistanta—put in da notebook—da froggie witha no legs, he-a deaf!'"

Emile laughed so hard it hurt. A joke about ethnicity, a joke about second-rateness. Dennis's own mind was first-rate, he'd known that since he was young, had never needed anyone's endorsement to know that.

He'd grown up with about four hundred or so other Jewish kids in a brick complex in Sheepshead Bay, one of the first co-ops in Brooklyn, colonized by middle-class couples after the war. The kids played all the New York games, punchball, stickball, slapball. The stickball strike zone was painted on the brick wall right outside Dennis's first-floor window. When they called Dennis out to play he came out his window.

His older brother Jay was their father's son. Sidney Priven was more sociable than his wife Mimi, more of this world. He worked as a printer, he led Cub Scout meetings in the basement. He did not take things as seriously as Mimi did, didn't see the hidden layers of things. Miriam Priven was more sensitive and intelligent, more contained. Some said she was negative. She played mah-jongg with the other young mothers all day long, liked to play so much she tied her second son to a tree by a rope around the middle to keep him out of her hair.

Dennis was like his mother, played out games in his mind. But Mimi was sickly, and became debilitated, even as the boy grew chunky and strong.

Other boys called him "the Ox," though he always had a sweetness. When the older boys tormented the younger boys, Dennis did it but his heart wasn't in it. But watch out when he lost his temper. Ray Fruchter came down the base path in punchball and stepped on Dennis's foot by accident, and Dennis screamed and chased him around the lot, and everyone moved out of the way. It was like Dennis was going to kill Ray, till Ray clocked him, that was the only thing to do with him then.

The Jewish kids feared the tougher Irish and Italian gangs along the Belt Parkway, but when fights came, Dennis stood his ground, and boys said he was afraid of nothing. He ran straight into a fence chasing a ball, bent the metal crossbar with his head, got up, and said, Let's keep playing.

He didn't seek recognition. Publicity went against his nature. Close friends knew that he was smart, more than smart—very, very bright, they said, deeply gifted in math. He had natural confidence, he didn't try to impress the science teachers at Sheepshead Bay High School or, later, his professors at Brooklyn College. He seemed to have an instant grasp of numerical concepts. He could get A's in organic chemistry or math after doing nothing all semester but looking over the textbook for a couple of hours the night before.

Girls were a problem. It was fine if he could treat them like guys but if he liked a girl he didn't know what to say. He'd fixate on a girl, and shyness would overtake him. He told a good college friend he was interested in a girl the friend knew, and the friend encouraged him, said, Great, talk to her. But when Dennis opened his heart, something happened. It went badly, it was suddenly painful. And after that he cut off his friend.

He was comfortable with men. He joined a college fraternity of ethnic kids, Tau Epsilon Phi, and played cards all night, and drank and hung out. He didn't seem to have a happy home life; it did not surprise his friends when they heard he was going to the other end of the world. Dennis was game, he'd do anything.

Now here he was, and Lampshade Phil saw his bearish charm and made him a charter member of the Royal Nuku'alofa Martini Club.

Tupou High was a low-slung set of buildings on the east edge of the downtown. It was a Wesleyan school, the words "Fear God" were on the wall of the assembly hall, and every morning before class the students and staff gathered for a religious meeting, hymns, and prayers.

Dennis avoided assembly. He hadn't come out to Tonga to sing hymns, he'd brought out his knowledge to help the students get into university. Some of the girls were afraid of him and saw fierceness in his eyes when he shouted at them for talking. But he made a connection to the best students even when he only grunted approval.

Keith Moala felt love for Dennis because he was the first teacher who could make math come alive. He chalked a trig mnemonic on the blackboard: SOH CAH TOA—

Sine equals Opposite over Hypotenuse.
Cosine equals Adjacent over Hypotenuse.
Tangent equals Opposite over Adjacent.

Then the math test was too easy, and Keith finished early and spent the last half hour drawing sausagey letters on the bottom of the page, BLOODY TEST, with drops of blood coming off the letters, and didn't even know "bloody" was a swear word, had just seen it in karate movies that came to the Finau Theatre.

Of course, Keith got the highest grade, but when Dennis handed back the papers, he locked eyes with him and held that gaze even as he sat down at his desk. He said not a word, but the expression said it all: You have broken the rules. You are not to break rules. I respect you for breaking rules.

How could Dennis respect the Tongan rules? Tonga was pagan and backward and Christian and hypocritical, and feudal too—they had noblemen here. He did not like the fact that when someone tripped and fell hard at rugby, the crowd roared with laughter. Or that one day in class he had just begun to talk about the sex organs of some animal when the classroom broke out in cries and shouts, students got to their feet.

"*Tapu, tapu!*"

Three members of the Prescott family were there, two girls and a boy. It was *tapu* for siblings or cousins to hear anything about sex in one another's company. So the class stopped as the girls were excused, and afterward in the Tonga Club Dennis said *tapu* was superstition from the Dark Ages.

"You're not being very culturally sensitive," said Berl, a volunteer from Alabama, but Dennis laughed in his sarcastic New York way.

"Who's not culturally sensitive? I lie, cheat, and steal, just like all the Tongans."

The Tonga 11s didn't care for him. Westerners and Midwesterners, by and large, with Protestant upbringings, they were good-natured men of a religious bent, struggling not to take themselves seriously and to learn from their hosts. "Hey Dennis—knock it off," they would say when Dennis got aggressive in the Monday night basketball games at the Mormon gym. Sometimes he came to the poker games at the mosquito lab and cleaned the 11s out. Mike Braisted walked away with three or four dollars gone, a lot of money, and kicked himself for letting his thoughts wander during the game. Dennis had been completely engaged. He could count cards, and didn't drink when he was playing, and seemed to know what everyone had been dealt.

"Why'd you play the jack of clubs when you were holding the queen of hearts?" he said to Frank at the end of the hand. He had second sight about cards.

The village of Longolongo, at the southwest corner of the city, was crisscrossed by grassy lanes of the sort that Captain Cook had described 200 years before, with natural fences that gave the village a mazelike feeling.

Dennis lived in a tin-roofed hut at the north side of the village. A couple of hundred yards along those lanes brought him to the Bird Lady, and another couple hundred brought him to a small cinderblock house, the McMaths'.

John McMath was deputy principal at Tupou High, craggy, thoughtful, good-natured, poetic. His wife Marie, pronounced MAH-ree, in the Australian way, was bubbly, with a strong rilling voice, beautiful white skin, and pretty dark curling hair. The couple had four towheaded boys, and bicycled through Nuku'alofa in a line.

The McMaths didn't advertise it, but they were Baptist missionaries. John had reached a point in his career in Perth where competition had started to wring the joy from teaching, and when he and Marie saw the ad for the job in Tonga, they prayed to God to know what he wanted from them. They were generous and saw the good in people. When word passed among the expats that a *pālangi* family had to send their child back to New Zealand for medical treatment, John showed up at their door. "We know you'd do the same if we were in trouble," he said, and walked away, and Laureen Munro opened the envelope at her kitchen table and wept. One hundred pa'anga—a lot of money, close to $140 U.S.

Marie McMath was like a sister to Dennis, she was affectionate and crossed the borders in him that others were afraid to approach. One day she even shouldered him hard in her kitchen and he fell to the floor.

"Dennis Priven, I am going to humanize you, you're going to be a little western Aussie by the time I finish with you."

Because he was such a New Yorker, like a cat on a hot tin roof all that first year, Marie thought. Taut as a bowstring, said John.

Love is subtle. The McMaths saw Dennis's sweetness, and Dennis felt comfortable in their house. When a Brooklyn friend sent him a snowflake in a little plastic vial, Dennis brought the drop to the McMaths and went on about New York and laughed hard, and that was Dennis at his best, carefree. But when Dennis was puffed up and masculine, and got out the Seahorse in the kitchen and talked about the blade's sharpness, and cut his thumb—Marie told him she didn't want him to do that in front of her children. That Seahorse went everywhere as some sort of security measure. He even laid it in its sheath on the blackboard sill, ready to fight off intruders to chemistry.

"Dennis, I don't want you taking that knife out in my house," Marie said sharply.

Still, he could relax with Marie, he could complain. "It's all right for you, you've got something to do on Sunday. I get bored."

"What's wrong with you? You can come to church."

"I'm a Jew."

"Well, congratulations—so am I. I'm a Jew by the grace of God, by adoption."

Because if you accepted the New Testament, she said, Christianity was the religion that God had given to the Jews. It superseded Judaism, so a Christian was a true Jew.

"Well, I'm a Jew by birth."

"So what! You worship God in your head, you can come along."

"No."

She felt that Dennis was a little embarrassed about being Jewish, and she said, "Oh, come on, Dennis, why can't you come to church?"

But Dennis merely shook his head, backed up in his chair, and Marie knew she was on a touchy subject. That was how he ended discussions, put the subject in a box and firmly closed the lid, or said, "It's a long story" and shut up.

Other Tonga 14s settled in.

Tom became Tomasi. His Om symbol fell off his neck into the ocean and he spoke Tongan so well he threw around masturbation jokes with the Tongan women and moved to an outer island, and later abandoned celibacy. The Bird Lady traveled everywhere making observations, sending them home. "I'm planning to count the number of normal & frizzy chickens & have my students figure out the heterozygous using ye olde Hardy-Weinberg formula (for changes in gene frequencies in time; evolution), but I'll tell them it's a one-gene simple dominant, though I'm not sure it is."

Emile became 'Emili, and danced for the King. On the 100th anniversary of the Constitution, the village of Ngele'ia put on a dance and Emile practiced for weeks and wore a grass skirt and anklets and a headdress, and the villagers pushed him out front on the field, proud of this tall graceful *pālangi*. A lot of volunteer friends came, but not Dennis. He didn't approve of that ritualistic stuff.

The place did not agree with him. He got stomach pains so severe he doubled over on the basketball court, or in Lampshade's martini club, and took a minute to recover. When a volunteer friend, John Sheehan, got married to a Tongan woman, there was a drunken bachelor party, and Dennis bumped into the groom and said, "Well, John, now you're the king of the niggers," and John took a swing at him. He began wearing jeans cutoffs and an old T-shirt to school, in defiance of the dress code for teachers. He threw students out of class when they didn't do their homework.

He went in and out of others' rules the way a bicycle passes in and out of shadow, and his friends admired that, how independent he was of authority. The teacher Gay Roberts's spirits fell every Saturday when she saw the plane come from Auckland bringing letters from her family, letters that would be locked away in the Tupou High School mailbox at the post office and all dreary Sunday would drip by when she could be writing home. Then one day Gay came into the teachers' staffroom and Dennis said, "I think this might work," and handed her a microscope box key he had filed down to fit the mailbox.

Well, that changed Gay's life.

He was like a brother to the Bird Lady, too. He designed the curve on her tests, fixed her bicycle, dug out her water pipe, and set up an experiment of agar blocks made with phenolphthalein. She wrote home about him in her taxonomical way.

"Good at math, chem, & succeeds at what he wishes to do. Since he has a beard & usually wears cut-off blue jeans, the Tongans think he's sloppy—which he isn't. Keeps his desk, bookshelves, home very nearly neat as a pin. The students are scared of him, not knowing that beneath that gruff exterior lies a tender heart of the sort that rescues fair damsels in distress. He'd hate to think so, though, disliking sentimentality.

"All in all he's too good to waste—I keep wanting to match him up with some fluffy little wisp of a girl with a will of iron. They'd live happily ever after."

At the end of his first year, Dennis was medevac'd to Hawaii for the mysterious stomach pains, and he flew back out with Deb's group.

In his bag was an Aloha shirt for Lampshade Phil and, in disregard of the obscenity laws, something to blow Emile's mind, a new magazine called *Hustler*, with a porn star on the cover named Long Dong Silver. Emile had to shake his head. "What has happened to the country since I left?"

Dennis's schedule in the second year gave him a half day on Thursdays, and every Thursday noon he went to the McMaths' house to do his laundry. They had a washing machine that rolled across the kitchen to the sink and plugged into an electric socket, and he sat in the kitchen with Marie as she prepared dinner.

He talked about systems. If you studied any system long enough, you could crack it. He'd been thrown out of three casinos in Las Vegas because he'd beaten the house so consistently. Marie was not sure if he was telling the truth, but she forgave him, he was 24. And likewise when he told her about a Brooklyn friend of his in a special agency a lot like the CIA but not the CIA, more secret. A phone call would come at dinner and he would leave the house, not come back for a few weeks.

Dennis invoked secret codes himself.

"Do your father and mother go to synagogue?" Marie said.

"That's on a need-to-know basis and you don't need to know," Dennis said. But with a hint of a smile, so Marie felt she was breaking through.

•   •   •

His friendship with Emile weakened in the second year. Dennis hated it when anyone played games with him, and Emile's humor sometimes hid his meaning. Maybe any friend would have failed Dennis's test. But Dennis made friends in Deb's group.

Most of the new volunteers tended to avoid him, he was too intense. Wichita wanted to scooch down a couple seats when Dennis wound up next to him at the Tonga Club. But Wichita was gone, and Paul Boucher and Vickie Redpath were East Coast people, smart and purposeful, they were open to Dennis's rude charm and thought it fine he didn't try and speak Tongan or wear a woven skirt. Dennis played bridge at Paul and Laura's house with Paul Zenker, a shy brainiac considered the smartest kid in Deb's group, a curly-haired bespectacled bio major from Notre Dame, an introvert, too, drawn to Dennis's intelligence. Dennis helped the Johnsons build a bed, and he did some of the Johnsons' blue microdot.

Then that July, the friends helped Dennis engineer his dinner date with Deb and Deb got angry about it. She felt set up, she said it had gotten ugly. Vickie Redpath felt chastened, and in her blunt way, she told the others that what they'd done was wrong, they must not do it again.

Strange news went out on the coconut wireless: Dennis had cut his door in half, making a kind of Dutch door, and painted a menacing face on his door.

The Bird Lady heard about it from her neighbor 'Uini. 'Uini said Dennis had done it to scare away *tēvolos*, devils in Tongan. A neighbor had died of dengue fever and then returned to Dennis's house after her death and begun disturbing his sleep. He had painted the face on the door to scare the spirits away, sort of a skull and crossbones, and also tried to appease the *tēvolo*, by setting out food for her.

The Bird Lady quizzed 'Uini, and 'Uini responded gravely, as if the visions had a meaning that could be logically unfolded. But when the Bird Lady brought the matter up with Dennis at school, he laughed and looked sheepish, said he had done it to keep prowlers away. Someone had broken into his house and taken 70 pa'anga, a lot of money to a volunteer who only made that much in a month, even if it was poker winnings from other volunteers.

"Mr. Priven, do you believe in God?"

Keith Moala questioned his beloved teacher after class, and saw that blank expression, as Dennis measured his sincerity.

"I believe that God exists. But I do not believe in him."

The statement unsettled Keith, and he puzzled the words like a math problem till he came up with a favorable reading. In Mark, Jesus approaches a crowd of people who are disbelievers, and a man comes out of the crowd to ask Jesus to heal his son who is possessed by an unclean spirit. Jesus says to the man that if only he believes, anything is possible, his son can be saved. Immediately the father cried out to Jesus, Help thou my unbelief!

That was what Dennis was saying. He'd not yet achieved faith, but he had accepted God's existence, and that was a first step.

"I have four main friends back in New York, men I grew up with and can trust my life to—"

Marie felt that Dennis's mind was going back to New York in his second year, trying to adjust to the idea of returning, and she teased him, calling his friends "The Big Four."

"They are my only real friends."

Marie's feelings were hurt. "What about us?"

"Well, it's not the same."

As if they had parted some tapa-cloth curtain to find themselves in another reality.

Still, he came every Thursday, rolled the washing machine across the kitchen to the sink and dumped his clothes in, and Marie cut fat away from the second-rate meat that was sent out from New Zealand.

"You making food reminds me of my mother making food," he said with a sudden motion of tenderness.

He felt anger toward his father but pitied his mother. He said he got upset with himself when he made his mother scared. Marie did not push the subject, Dennis did not elaborate. Still, she didn't think that he had scared his mother by going up behind her and saying Boo. It was something more.

7

Stranded

Deb wanted out. She told Judy she hated Nuku'alofa and told her mother she was lonely. "I'm feeling stranded—"

She took a vacation from the city with three other volunteers to deserted islands in Ha'apai, Tonga's middle island group. There was nothing living in the bush on 'Aku 'Aka but one lone pig, and the fisherman who brought them out did not want to leave the *pālangis*, he worried about water, but the volunteers insisted. For the first couple of days no one could get a fish. They sat at the fire roasting sweet potatoes, then Deb pulled a can of corned beef from her backpack, and chocolate.

Sargent Shriver had once made it a rule that volunteers confine their vacations to the Third World, and this is what he had in mind, Americans wandering with a bush knife and backpacks in the scattering of islands where the mutiny on the *Bounty* had begun in 1789. Just Deb and three guys, David McLean, Paul Zenker, and Jon Lindborg.

Jon was from a worldly family in Monticello, Indiana. His parents had boarded exchange students when he was a kid, and now he was becoming *faka-tonga*. He was strong and sweet and marveled at Deb Gardner. She didn't complain, she knocked down coconuts for water and husked them with a bush knife. She dragged the long dried ribs of coconut leaves from the bush to build a fire and swam out past the reef wearing a WSU shirt and holding a speargun she wasn't good with. Were all girls in the Pacific Northwest like that?

Jon had heard her reputation, Deb Gardner is the siren of the Peace Corps. Now he saw a small-town kid like himself, some kind of frontier girl. Zenker and McLean had a farting contest by the fire. "Listen—you can hear the fish sing," Zenker said, and everyone shut up and he let one rip. Deb laughed. She was innocent.

After the second deserted island, Deb and Jon got a boat to Lifuka, the main island of the Ha'apai group. They met the principal of Tāufa'āhau Pilolevu High School, and he asked if they could come teach there the second year. Mark Stiffler was transfering to Ha'apai, too—tall blond Stiffler from Sonoma County, who was determined not to be the Ugly American and who could spend all Saturday in the water and all Sunday in church.

Deb was determined to go. Mark and Jon weren't hitting on her, they were her friends. Though, no, Jon couldn't help but think about Deb in that way. On the boat back to the Big Nuke, she got cabin biscuits and corned beef from her providential backpack and as she spread the stuff out said, "This stuff always reminds me of dogfood." They saw humpback whales breaching at sunset, then Jon stretched out on his back and Deb pillowed his head in her lap. It felt pure to him, like a high school crush.

Vic Casale had told Mark and Jon to watch out for Deb when he left Tonga.

Having failed two or three times to get a medevac from Mary George, it finally occurred to Vic that he'd erred by listing all his problems to the director. She was beautiful and self-involved; he had to woo her to get his medevac. So he made an appointment to visit her at her house on a Saturday.

The director's house was a big place on the foreshore a mile or so west of the palace. It was constructed in a neo-*fale* design, with oval ends echoing the lines of a coconut house, but Tongans derided its location, a few steps from the sea, lonely and lashed by storm. Vic showed up in his white pants and pressed white shirt, and the two of them could have been in Georgetown. Mary poured Vic a drink and he sat on the couch and looked out the glass patio doors at the gardens, and Caesar, Mary's power cat, a gray Siamese, wound between Vic's legs.

He asked her all about her travels. She'd left Maryland at sixteen to become a fashion model in New York and had stayed in the business till she was almost thirty. She'd made as much as $100 a minute. She had been in the Mennen ads that ran with the fights on Monday nights.

She'd lived in Paris, she'd studied French. She had worked as an executive secretary in Washington, and run the Barbizon School too. But the hours were crazy. She'd worked night and day for nine months and at the time she was a single mother with responsibility for her daughter.

What a coincidence—Vic's friend Susan had gone to the Barbizon School. "You might know her, Susan's dad was vice president of Wrigley's chewing gum."

He was careful to say "my friend Susan." He wasn't about to say Susan was a former girlfriend. Mary didn't want to hear about other women when she was sitting across from him on a couch. Vic had figured out that the deal was not to rattle Mary. She was an Audrey Hepburn type, with fine bones and large commanding eyes, fineness in her every pore, and she was highstrung, she got cross easily.

That was the only rough spot in the conversation, when Deb's name came up.

"I thought you were going out with that Debbie Gardner," Mary said.

Vic deflected that. "Deb is OK, she's all right. Everything's cool—"

He gave nothing away, but that threw him. Who was Mary George to ask about his relationship to Deb Gardner? That was none of the government's business. Deb had moved on, he'd moved on. Still, he didn't like the look in Mary's eyes. She'd flashed disapproval, or jealousy. So Deb had managed to piss Mary off. That didn't surprise him. Deb was headstrong, natural, and naive, while Mary was sophisticated, and maybe a little tough.

When he came out of Mary's house later that afternoon, Vic had his medevac, and for a long time after that other volunteers talked about how he had dressed up for visitin' to get his ticket home.

Mark and Jon took Vic out to the airport. "Look out for Deb, OK, guys—keep all the animals off her," Vic said.

He never did give Deb a proper goodbye. Maybe there was a little bitterness, a sense of what might have been, but when he got his mail for the last time and came down the stairs by the lagoon, Vic saw Deb sliding off her black bicycle proudly, saw the hurt in her eyes. It was his move, and he should have said something. But Victor Casale who had quicksilver instincts about women actually faltered for once, and then she turned out of his life.

After Deb got back from Ha'apai, all she could talk about was moving there. She bicycled with Mike Braisted out to the Mormon high school to hear a violin concert, and told Mike she really needed to get to an outer island. She hadn't felt hemmed in there, she felt confident and independent. Then Frank came back from leave, and his marriage was over now, forever, but Deb said she wasn't getting serious about anyone now, she just wanted to get out of Nuku'alofa.

She went to see Mary about a transfer, and Mary told her she wasn't going to put any more single women in the outer islands. There was too much risk for them being out there by themselves, away from the Peace Corps administration. Look at all the problems Judy was having in Vava'u. Men came into her house in the middle of the night, stood at the end of her bed to ask her to marry them.

There was a tone of judgment in Mary's voice. Mary seemed to suggest that Deb had a reputation.

Well, that was unavoidable. Nuku'alofa was a crowded village. Tongans called gossip the kerosene radio because it provided entertainment but it always sputtered and stopped just when it was getting interesting. You didn't know for sure. Older women said that Debbie Gardner was free as the air, that a man could have her as easily as he took a breath of air. Two New Zealand mothers were pushing their strollers in the King's Road when Deb rode past, hair flying, and one turned to the other. "That girl is going to come to a sticky end."

One night Deb was sitting around with Francis and Vickie and an English friend of Mary's called Tanya when Tanya said, "Women shouldn't go out by themselves at night. They shouldn't even ride their bikes by themselves."

"I'll ride wherever I want whenever I want," Deb said.

Tanya was patrician, and she said, "But in a foreign country you have to respect the different mores. A woman by herself in Tonga is seen as somehow deserving what happens to her."

Deb wanted no part of that. Women had so little freedom in Tonga. Everybody assumed the worst of them when they were out on their own, while men could go out whenever they wanted for any length.

"I'd rather die than not be able to go where I want when I want," she said.

Mary George had her own reputation. It was impossible to be a single attractive woman in Nuku'alofa without getting one. People said that Mary was overly attuned to men. She would touch a man on the upper arm and his back and gush and make him feel special. If a man came in the room, Mary would cut a woman right off to go up to him.

She kept male company. First there was a guy from the State Department. Then he was gone and an Englishman called Fulcher came out to run a shipping line. He was tall and angular, stiff, handsome, fatherly, with steel-dark hair, and Mary and George Fulcher were seen everywhere. They went to dress-up dinners, they went boating together, they got caught out in the rain together in their good clothes, and Mary ducked into George's room at the International Dateline Hotel to change her wet clothes, and a man banged on the door shouting, "No guests in the room," a policy the hotel maintained on account of the *fokisis* gathered across the street on Yellow Pier, and George went to the manager in anger.

"People will think one thing of her, that she's in there on her back. Now you must apologize to Mary."

But who could stop the kerosene radio?

George Fulcher said that the relationship was friendly, there was a girl in Australia he was going to marry. Besides, Mary was political to her eyeballs and George didn't care for politics. Mary could go on and on about a dinner she had gone to at the White House—

"George Balanchine was there and Ernie Ford. Tennessee Ernie Ford? He sang his coal mining song. 'Sixteen tons, what do you get—' And Elke Sommer was there with her husband, and Angela Lansbury. And everywhere you looked was someone you recognized from television, Ethel Kennedy and Dick Cavett."

What did George Fulcher from the Liverpool docks care about Elke Sommer? He didn't.

◆   ◆   ◆

Peace Corps country directors were warned during their training that because volunteers made no money and were dubious of authority, they could turn on their directors, and curse their fancy houses and their $28,000 a year. Now the volunteers turned on Mary.

The Volunteer Council idea became a monster because Mary did what she wanted and didn't really need anyone's advice. "You never listen," a council member said, and Mary said, "I always listen to the first couple of sentences you say."

The volunteers told mean stories about her. She couldn't walk on the beach barefoot because she was so used to walking on heels. She had a special box with a lightbulb in it to keep her shoes from mildewing. On a boat trip one day, she kept yanking the bottom of her bathing suit down then asked a volunteer, a 25-year-old girl, how she looked.

She expressed keen sympathy for all the miserable and injured dogs in Nuku'alofa, and asked if there wasn't some way for Peace Corps to bring in dogfood for them, from New Zealand. Good grief, could you imagine the poster for that one back in Washington?

She was withering to some volunteers, while others she killed with kindness. The Bird Lady went into the Peace Corps library to find a book—

"Our new director smiled a greeting," she wrote to her parents. "I reluctantly returned it. Next thing I knew she was bearing down on me. No place to escape. She was so happy to see me. She hadn't known I was interested in birds until just last week when someone told her. How wonderful, she was interested in birds, too! Why did I never come to the office? Here I firmly interjected that I rarely came, only to get my mail, & it didn't come often. She understood! She didn't get much mail—well, only business—and on Saturdays her daughter wrote. How wonderful Saturdays were.

"She continued on the subject of birds. When she deplored the total lack of reference works on Tongan birds I managed to tell her about Ed Carlson's work & that it was available in her assistant's office. She didn't seem fascinated. She'd gone to the Audubon Society before she came and they'd had nothing, just nothing on this area.

"So it seems our new and unbeloved country director is a bird freak. A gushy one, no less. Perhaps I will turn to entomology."

Peace Corps/Washington had known about Mary's liabilities. One staffer even listed them: lack of self-control, insincerity, losing her temper. She worked extremely hard but was dramatic and tended to burn out, and burn out those around her. Her main professional experience after the age of 30 was at the elbow of one powerful man or another. She'd worked as an executive secretary to an admiral, confidential assistant to a former ambassador, girl Friday to the American ambassador at the Paris Air Show.

It wasn't as though Peace Corps had much choice in the matter. The position of country director was a political appointment, at the pleasure of the president. By and large Peace Corps people were able to control this process, and ensure that the great majority of the sixty-three directors had experience in management, the Third World, antipoverty work.

But a few of the appointments were frankly political. "Must place"—the White House insisted on the appointment. Sometimes the White House even sugared off a big donor by telling him that being a CD was like being an ambassador. Then the man would cross Lafayette Park to the Peace Corps office and ask how they were going to ship his Mercedes out to Bamako, and the Peace Corps had to explain that driving a Mercedes was a sure way to alienate volunteers who had taken vows of poverty to live on root crops and eggs.

Mary was a must-place. She'd worked Capitol Hill as a lobbyist for the Export-Import Bank and grown friendly with an old friend of President Ford's from Grand Rapids, Jack Stiles. In times gone by, the President had sat up with Stiles on Election Night to watch the returns. Now Stiles was a White House counselor, and he took Mary to that presidential dinner and for nearly a year after that her name was passed around at the personnel office, till the idea of Peace Corps/Tonga presented itself, and people said there was no freedom or challenge quite like running a Peace Corps office.

She didn't bring a Mercedes out to Nuku'alofa, just Caesar and all those shoes. And meanwhile the answer Peace Corps came up with for Mary was the answer it came up with for any problem, "Find a great volunteer, throw him at it."

Michael Basile was Mary's assistant director and the only other *pālangi* on the staff, 35 years old, a lean, cerebral, reserved man with a triangular face, dark heavy-lidded eyes, and Peace Corps values. Mike had been a volunteer himself in Turkey in the 1960s, then worked on the Peace Corps staff in Iran and in the antipoverty movement in Washington. His wife Janice had been a "supervol," one of those rare Peace Corps volunteers who in addition to finding out who they were actually accomplished a lot. In Brazil she'd gotten schools built.

From the start there had been strain between Mike and Mary. Mary was overpowering. Mike tended to withdraw. On vacation in Fiji he ran into a headquarters official named Hailer and confided in him.

"We're having trouble in Tonga. Mary is always in crisis. She turns small disagreements with volunteers into major battles. Someone in Washington really ought to look in on this."

Hailer said he would see what he could do, but when Mike got back to Tonga, Mary called him into her office.

"Did you say something to Dick Hailer about problems in Tonga?"

"I just told him some things that were on my mind."

Mary got very upset, said Mike was talking out of school. She didn't want anything going back to Washington by a back channel.

"You have to promise me that you will never talk about me behind my back in this way—"

So Mike promised.

Dennis had a new plan, to stay on for a third year. Dennis re-upping for a third year! That was madness. Dennis did not like Tonga, why was he trying to stick around for a third year? Some volunteers were afraid to go home. They fit in better here than in the States.

Kalapoli Paongo, the principal of Tupou High, did not want him a third year. Mary didn't want him either, and Mary tried to finesse Dennis. She would have to observe him in the classroom before she approved his staying on. But, oh how Dennis despised Mary, and felt she was incompetent. She wasn't allowed in his classroom, he said. She was not qualified to assess his work. And other volunteers supported him.

Dennis went to Mike Basile. Ordinarily, Mary dealt with all the teaching volunteers, and Mike with the technical volunteers. Dennis came in with a tense look on his face, lacked any ability to make small talk. And that knife on his belt.

"Mike, I come to you because I hear that you listen. This is very important to me."

He really wanted to re-up for a third year, he wanted to go to Ha'apai of all places. . . .

"Would you speak on my behalf to Mary?"

Mike agreed to do so, and Mary yelled at him.

"He's not qualified to extend here and we can't lead him on! Mike, you've got to take a stand in this job sometime. You always leave me with the tough nos. You're good at saying yes all the time. I need more support than this out of you."

Dennis was focused on Deb. "What do you think if I put a listening device in Deb's *fale*, a microphone so I can hear who is coming and going?" he said to Emile, and Emile said, "This isn't Russia, Dennis." Then another time Dennis said, "I know what I'll do. She loves that cat, I'll kidnap her cat without her knowing it. Then when she's looking for it I will be the one to find it and bring it to her, and she will be very grateful to me."

Emile shook his head. "I don't think so."

It was September. Mary was in her corner office, using a can of Raid to zap all the insects that came through the woodwork, when there was a soft knock on the door.

She looked up and there was Emile. "Emile, come in," she said brightly.

"I need to speak with you about something important."

"Sit down, close the door."

"It's about Dennis—"

"Yes?"

"He seems to be changing. He's gotten depressed or withdrawn, he's hung up on Debbie, he's said some pretty strange things to me—"

Emile stumbled, not being used to such serious errands, and Mary got a tight smile, and he was not sure how far to take it.

"Dennis only has a few months to go," Mary said.

"He seems to be slipping off the rails—"

But Mary said that every volunteer got a little down as the months dragged on and the end date didn't seem closer, and they also didn't know what they were going to do when they got home. "Dennis just has island fever," she said.

Emile realized that Mary had a lot on her hands, he was being dismissed.

A fresh group of volunteers arrived at the end of September, the first group that was all Mary's, and she drove out to the airport in her Land Cruiser followed by a bus with a few older volunteers on it who were serving as a welcoming committee.

It was a hot and dusty day. Deb wore a mannish outfit, faded olive khakis and her bandanna around her head, coral beads around her neck, and Emile wore a tapa-pattern shirt and church pants, scratchy wool bell-bottoms, gray with a red stripe, that seemed to have been abandoned by some nineteenth-century missionary.

Tonga 17 was a technical group, engineers, writers, public health experts, phone technicians, and they were blown away by this vision in pale blue with a matching scarf and heels who had champagne for them in the VIP area and seemed to have studied up on them. Mary knew that Bob Forbes played guitar. Knew that Jan Worth was coming out to do public relations for the Kingdom. Knew Rick Nathanson had been a reporter all through college.

She sped into town ahead of the crowded bus, trailing a plume of coral dust, her scarf flapping wildly out the window.

There was trouble in the new group. Rick Nathanson was a short kid from Chicago with wild black hair and a forthright style, but his FBI background check had turned up the fact that he'd gone to events held by a socialist organization at the University of Illinois. Famously, three things could get you thrown out of Peace Corps, the three P's: politics, pot, and perversity. "I'm not really a socialist, that was just the only place to meet cute Jewish chicks," Rick explained, but Peace Corps had him on probation.

Mary took him aside at the office. "Now Rick, you won't be getting involved with politics, will you?"

"I'm going to work for the government newspaper, how am I going to avoid getting involved in politics?"

"There are lots of fun things to write about here that don't involve politics."

"But isn't that a little incongruous? I mean I'm working for the prime minister's office. It's a political job."

Mary kept at it, and finally Rick nodded and agreed. "You know what, Mary, you're right."

Rick had come out to work for the *Chronicle*, and unlike Frank, who'd fallen into journalism, he was well-trained. That was something the King had specifically asked for, American journalists to bring first-world standards to the Kingdom.

Paua, the editor of the *Chronicle*, had gone out to the airport for the arrival to take pictures of Rick and the other volunteers. He got one of Deb as she waited for the plane.

Two men were clowning with Deb. Emile draped his elbow on her right shoulder like he was balancing against a wall and pointed at her with his free hand as if to say, Look at her! A third-year volunteer named Rich Dann was on her left. He had one arm around her and his nose against her neck, almost as if he was trying to look down her shirt at her coral beads. She smiled and put her arm around Rich's back, but kept her fingers closed in a fist.

The picture went up on the Peace Corps office wall.

8

The Dance at the Dateline

In October, Dennis went by Tonga High School every morning, and Deb hid from him in the book room and told her friend Talahiva, another teacher, to say that she was not there. So Talahiva did that for Deb, lied to Dennis, and one day the *pālangi* vice principal confronted Dennis.

"This is not your school, you teach at Tupou High, you should not come here in the morning."

In the afternoon at four or so, Dennis showed up at the Peace Corps mailroom and waited for Deb to get her mail.

The Tongan staff members did not care for him. Makaleta (Margaret in Tongan) was the office manager, Deb had shared her Sears catalog with her, and Makaleta could see that Deb was afraid of Dennis. Other staff members noticed Dennis's knife, wondered, Why does he carry that big knife?

On October 5 the Ambassador came out from New Zealand, and Mary staged a dinner for him at the office. She invited all the volunteers, but as there wasn't enough money in the Peace Corps budget for canapés and prawns, she said that it was potluck and asked the volunteers to spend 10 pa'anga apiece on food. Ten pa'anga was about $13, real money if you made only 70 pa'anga a month, and as a form of protest, Deb and Emile came with coconuts. "Sometimes we have coconuts for dinner," they pointed out to each other, "the ambassador can have coconuts too," and they put three or four of them together with a ribbon and Emile called it a coconut bouquet.

It was very funny to them, pulling the coconuts out of Deb's backpack, putting their thumbs in the plutocrat's eye. Deb got giggly over it.

Mary went everywhere at the Ambassador's elbow, called him Mr. Ambassador. He was a Southerner like herself, a genial pear-shaped man with his dark hair slicked over his round head. The volunteers all got to shake the Ambassador's hand, and the Ambassador made a joke about the Russians following him everywhere around the Pacific. Because at this time, the Tongans were flirting with the Russians, offering to allow them the use of the deep port in Vava'u for their fishing ships in exchange for fish.

Rick Nathanson asked the Ambassador if he could have an interview, and the Ambassador told him to come by the International Dateline Hotel the next day.

The Ambassador's door opened on steam. "Come in, come in—" He was getting dressed for a formal dinner, had a towel on but that seemed to be all, as if they were in some locker room back on Capitol Hill. Rick asked him about the Russians, and the Ambassador stopped and considered.

"As far as I know, the Soviet ambassador has been here only once—to present his credentials. . . . I have been here many times."

Then he dropped the towel and reached for his boxer shorts, and Rick went out feeling proud of himself. *He* had interviewed the Ambassador in his hotel room, *he* had seen the Ambassador's balls. How many 23-year-old journalists could say that? None. None but Rick Nathanson from Chicago, who had come to Nuku'alofa to write for the best damn paper in the South Pacific.

The Ambassador had one important piece of business in the Kingdom: to introduce the new chargé d'affaires.

The Ambassador lived in a house in a botanical garden in Wellington, where he met up with visiting American dignitaries like Vice President Nelson Rockefeller and the vice president's brother David Rockefeller. His diplomatic portfolio included Fiji, Samoa, and Tonga, but he rarely got out to the islands and as his name was Armistead Selden, the volunteers called him the Ambassador Selden Seen. If there was heavy lifting to be done in the tropics, it fell to the chargé d'affaires: the man who ran the American consulate in Suva, Fiji, about 500 miles across the water from Nuku'alofa.

The new chargé d'affaires was a lean silver-haired man with a very cool expression. Robert Flanegin had worked on newspapers in Indiana when he was young, developing an elegant style, but his temperament was close-mouthed, dutiful. Over his long career, the State Department had moved him around as a courier and then a consul in many out-of-the-way places. Paramaribo, Port Louis, Suva, Nuku'alofa, Apia—Robert Flanegin was a walking, talking game of Name That Capital.

Now the Ambassador brought him around to meet all the important members of the Royal Family. The King. The King's brother, the prime minister, Prince Tu'ipelehake. The King's son, Crown Prince Tupouto'a, first secretary to the prime minister for foreign affairs.

Selden got on very well with the King. Like His Majesty, the Ambassador was a formal man who enjoyed dressing up, and he took great pleasure in explaining to other *pālangis* the mechanics of the kingly audience.

"You will be served champagne. The King will be served orange soda. Go ahead and drink the champagne, but you'll notice the King won't touch his orange soda."

Even as the soda sat there, His Majesty would be delightful company. He might explain his method of doing arithmetic, going from left to right—"Right to left is pandering to Arabic tastes," His Majesty said—and he might throw off some witticisms aimed at his own people. Just don't say that Tonga was a British colony. They're very sensitive about that, they've never been colonized, though, yes, for a time Tonga was a protected state of the British empire.

Then the King would lift the orange soda to his lips, the signal that the audience was over.

The American group moved around town in an entourage of Peace Corps Land Cruisers. Dave Wyler was at the wheel of one. The Tonga 11 had lately come into the office as a volunteer staff member.

"Do you want to see how volunteers live?" Mary said, and Flanegin said yes.

They stopped in at a little Peace Corps *fale* near the sea about halfway between the palace and Mary's house. By outward aspect, this house was a *fale-tonga*, it had a thatched roof and was sided by *kaho*, a bamboolike reed the diameter of a man's finger. The Ambassador stepped from the hut back into the sunlight with an irritated expression.

"Good God, you don't mean to tell me Americans really live there—"

"Well, yes, actually, they do," said Wyler.

Mary could be excused the mingy potluck for the Ambassador because on October 9, the following Saturday night, all the volunteers were invited to a feast and dance at the International Dateline Hotel to welcome the new volunteers.

The volunteers were soon drunk with food and wine. The Dateline served whole roasted pig and chicken and fish and clams and mashed potatoes and real roast beef, and there was still chicken left when they went for seconds, and thirds of ham and clams.

After that, fruit salad and a cake with rum on the bottom and custard on top.

As they ate, the Polynesian floor show started. Bare-chested men in tapa skirts did warlike dances, twirling clubs, and grinned more broadly than Tongans typically did. For it was Tongan custom that the men go smiling into war. Then a line of women padded out delicately from the back and danced mostly with their hands. Their shoulders and arms were shiny with coconut oil, and people who appreciated the performance crossed the dance floor and stuck pa'angas to the dancers' bodies, a customary tribute in the country, and the reason that so much of the money felt greasy.

There was an open bar, and volunteers who could not budget three or four Steinlagers a week at the Tonga Club got smashed, and when the floor show ended, they took to the dance floor.

The ballroom was not really a ballroom, it was open-air, a broad red terrazzo floor under the sky. Most of the tables were shoved back under a partial roof, but the dance floor was exposed and got slick from rain or dew.

Deb was drunk. She wore a white dress she had just made on a sewing machine at Tonga High, and to the new volunteers she was alluring. At her table she leaned back in her chair and lifted her shining leg in the air and said to her neighbor, "Let's see who can lift their foot over their head," with her dress slipping down past her knee.

The volunteer beside her was thrown, but Jan Worth, who had come out to write public relations, gazed at Deb with a kind of awe. She seemed the embodiment of femininity, unapologetically erotic. That is what so many saw in her, a free and sexual spirit, but Deb Gardner was besieged and miserable, a star who did not want to be one, and when she stood before the mirror in the Dateline bathroom and brushed out her hair, Kathy Sullivan from Dennis's group met her eyes and a dark thought entered her mind: Boy, is that Debbie Gardner gorgeous, and why is it that the lives of the beautiful are so often tragic?

Notwithstanding the warning from the deputy principal at Tonga High, Dennis had continued to ride into the courtyard in the mornings, and his muscular thighs stood out against the jeans shorts he wore and his knife hung from his belt and he didn't look anyone in the eye. The girls at Tonga High came up with a name for Dennis. The girls had seen a kung-fu movie with a villain called Lafong in it, so they called Dennis Lafong, or Lafong the Killer.

"I am scared of him when he comes. It is almost a sensory thing," Talahiva said to Deb.

"I know, but I don't have anything to do with him," Deb said. "I've never encouraged him, and he keeps coming."

The head girl prefect in form 6 (the equivalent of grade 11 in the United States) saw Dennis walking alongside Deb down the school's outside corridor, trying to talk to her in an intimate way. Dennis touched her arm. Deb flung her arm in the air to brush him off, and that came as a surprise to the head girl. Miss Gardner was always sunny, 'Asinate had never seen her be abrupt.

At the Peace Corps office as well, Makaleta understood that Deb had complained about Dennis. Makaleta said that Dennis was *namu peka*, he smelled like a bat. That is about the worst thing you could say about anyone in Tonga, because a bat sleeps upside down and its shit falls over its own body.

·   ·   ·

The first time Deb fell on the dance floor it might have been the slick terrazzo. She skittered away from her partner and tumbled laughing to the floor, then got slowly to her feet. But the second time she lost her balance, she gave into it and stretched out on her back and gazed at the stars.

Every volunteer got lying-down drunk at one time or another, every volunteer got *konā*. Many did that night. But across the ballroom, Deb was being judged. Mary's bright sharp eyes found Deb, Mary felt Deb was making a spectacle of herself.

Mary had special cause to worry about decorum that night, for she was once again at the elbow of a powerful older man. A day or two after the Ambassador and chargé d'affaires had departed, a high Peace Corps official had come out from Washington: the regional director.

Sargent Shriver had structured Peace Corps to resemble the State Department, and the regional directors were equivalent to assistant secretaries of state. Peace Corps divided its responsibilities among three regions: Africa, Latin America, and a third region that was no real region at all, but a long peel torn off the orange from Tunisia to Tonga and called NANEAP, for North Africa, Near East, Asia and Pacific.

NANEAP was Jack Andrews's realm. He was in his mid-fifties and handsome in a blocky Robert Mitchum sort of way, but Robert Mitchum just after he'd had a slug of castor oil. He wore a sour purposeful expression, untelling. He had been an Army Air Force lieutenant, later an executive in the department store business in Ohio, and then seemingly experienced a midlife crisis and applied to be a volunteer, but Peace Corps had had a better idea and given him NANEAP.

Jack was corporate, he wasn't Peace Corps. One day a woman pinched his behind in the elevator, and he turned to her with a look of grim triumph. "See—I am a tight-ass."

Now here he was in Tonga with a deputy, David Ingram. The two of them were on a routine tour. That, and checking in on a problem situation. The two men had barely arrived when Mike Basile took Ingram aside.

"David, I'm having real problems with Mary. She's created difficulties with volunteers that tend to end all communication with those volunteers, and it's hurting the program."

Ingram, a former volunteer himself, frowned. "Jack doesn't want to hear this." Jack believed in chain of command. If Mike was having troubles with Mary, he shouldn't jump rank, he should work it out with her.

Jack identified with his line managers. Mary was in an extremely lonely and difficult job, a single woman living by herself in the middle of nowhere, responsible for a bunch of young escape artists, assisted by only one other paid American, who had lost respect for her and was talking about her behind her back.

In an emergency Mary might phone headquarters, but phone calls were very hard to place. The call had to go through a switchboard operator, who then sent it on by shortwave radio to Fiji, where it went on to an oceanic cable. Washington being seven hours ahead of Nuku'alofa, people at headquarters were calling it a day when the Tongan switchboard operators got to work. The primary means of communication between Peace Corps/Tonga and Peace Corps/Washington was a cable machine, a telex that rattled the barn as it spat out a one-inch tape, perforated with holes.

The second time Deb fell down at the Dateline party, she ran into Mike Braisted, who wanted to dance, but she said she was going out to see the whales, then she and Emile walked away from the lights and the noise and out to the sea.

They walked past the loose knot of Tongan girls across from the hotel, past the reach of the music and the sharp cries of volunteers on the dance floor, and continued out Yellow Pier, a solid causeway the American sailors had built during World War II. It was a clear night. The moon was full. Deb wanted to go into the water, so they scooched down the big rocks that held the jetty above the sucking sea and Emile lowered his legs into the water, and Deb bunched her new dress around her knees and cried.

It didn't look like she was going to last more than a year in Peace Corps. Mary didn't seem to like her and was sure to deny her a transfer. Meantime she missed her family.

She'd gotten more letters about her dad. He'd come down from Anchorage to visit his own mother and father in Richland, Washington, and Craig had come over, and Grandma Agnes had overheard Craig and Wayne talking. Agnes had written Deb to push a reconciliation.

"Life is too short to have any ill feelings toward anyone," Agnes said. "It's hard to be young, it's also hard to be old. If the older people would try to understand young people, the young people would then understand their elders.

"I know, I haven't forgotten when I was young."

Maybe it was best to go home. Deb missed her dad. She didn't like the cold war, she wanted to talk things over with him.

Emile always tried to be positive. Everyone had rough times in Peace Corps, then you came around a corner and things looked up. If you just hang in there, you can make it work, he said.

Deb vomited quietly in the sea and some went on Emile's pant leg, and he washed it in the water, and after that they decided to sit in a boat. Boats were tied to the lee side of the wharf, their ropes moaned in the wind. Emile slid into the water up to his thighs and carried Deb to a boat and called her a princess.

From time to time a muffled sound came from the dance.

"Back to the dance or home?" Emile said.

"Home."

He went for the bicycles.

Rick Nathanson was so drunk that it came to him he had better leave now while he could still walk. He stumbled out of the Dateline and switched on his flashlight and tried to figure out the way to the guesthouse. He took a right and then a Tongan girl in a pretty dress came up from the shadow.

"Excuse me, will you walk me home?"

Rick wobbled, not believing his luck. "Sure—what the heck."

He tried to look at her as they went down the road, but when he lifted the flashlight she pushed it down. "No no no, I don't want my brothers to see me," she whispered, and afraid that he was breaking some *tapu*, that someone was about to jump from the bushes with a machete, Rick kept the flashlight down.

Emile walked by him going for the bicycles. He held Deb's bike by the handlebars and his own by the seat wheeling them to the foot of Yellow Pier. Deb was revived. She was talking about Ha'apai with two new volunteers, a married couple. Nuku'alofa is a big city next to those islands, she said. You climbed a little bent coconut tree to get water, you saw a thread of smoke going up from the volcano on the horizon. Flying fish jumped out of the sea by your boat, beating their fins, and at the end of the day a Tongan gave you a little *hopi* to drink.

She and Emile walked the bicycles home. They turned their backs on the sea and took the road down past the Tonga Club, slipped left into the bush on the Ngele'ia road.

Palm trees were blue in moonlight, and the Peace Corps *fales* made bright white boxes against the charcoal bush. The box on the left was bigger than the one on the right. Emile leaned his bike against his house, then followed Deb into her *fale*.

Two banana trees stood outside her door. One was tall and dark, with many tattered leaves, and beside it was a young one with broad untorn flaglike leaves. Dawn came inching over the ribs of the young banana leaf, and Emile stepped out the door.

9

Prayers in the Bedford Truck

His Majesty liked to say that his people were a very religious people, but he didn't know how many of them were Christians, and in the bush near the south coast was an old *pālangi* shacked up with a Tongan woman under a giant banyan tree who could explain the King's statement. "Tongans worship each other," he said.

The missionaries had brought Jesus Christ to the Kingdom just 150 years before, but the old beliefs were still alive: that godliness was in another person, and evil too, that a number of realities coexisted on the island at any time, that it was possible to move from one reality to another one, and maybe even from one time to another, by visiting the darkness or a bush doctor.

In the days before her death, a person was offered some premonition. Afterward the living could study that interval to try and find the sign.

So in days and years to come, the girls at Tonga High School reported that on Friday, October 8, Deborah Gardner put the finishing touches on the white dress she was making for the dance the next night at the Dateline and tried it on inside the home economics building, then walked through the doorway into the sun as several girls looked on. "Here comes the bride," she sang, mocking the song, and that act satisfied the Tongans' understanding. Wearing a white dress in the sunlight and thinking she was a bride when she was no earthly bride, that had been Miss Gardner's glimpse into the future.

Of course that was a quaint belief, and very Polynesian. But in this case it surely matched some more literal *pālangi* understanding.

Deb's death had been prefigured in several ways. There had been warnings that would only come out later, or not at all. *Pālangis* had heard threats. In weeks to come it was reported, quietly, that Dennis had actually threatened to hurt Deb a couple of times and that people had failed to take it seriously. Later, too, the principal, Kalapoli, would search Dennis's desk at Tupou High School and find a bunch of notes in the drawer between Dennis and a couple of other volunteers that seemed to Kalapoli to contain a rising level of menace, that appeared to refer to Deb as a pig.

Did Deb know anything about these threats? Of course she did. And she met them with a sort of desperate vivacity.

Nuku'alofa was like a big house. No one could really leave. A woman might escape someone else by going into different rooms, but that maneuver couldn't last too long. There was in the end no personal space. If someone wanted to find someone else, he could do so easily before too long. If a woman was afraid of someone, it was almost better to stay within sight of him in public.

So she did what she could under the circumstances, and went out with him.

On Monday night, a British astronomer gave a talk about crossing the Sahara, and Dennis asked Deb to go, and she went. They rode their bicycles through the western part of the city to 'Atenisi.

'Atenisi is Athens in Tongan, 'Atenisi was where foreign scholars came for a year or two, or sometimes a few weeks. The hall was crowded that night, the lecturer was a Welshman called J. F. James who had come out to Tonga for a lunar eclipse in ten days time. Not the actual eclipse—that could only be seen on the Indian Ocean—but J. F. James had come to observe the shadow of the moon at the end of the eclipse, as the shadow left Earth and passed through the airglow of the atmosphere.

The astronomer spoke about his travels. Three years before, he'd been to the Sahara for the solar eclipse, gone out by Land Rover across northern Africa. He told of the spot he'd camped at in southern Algeria that was the antipodes to

Tonga, which meant that if you drove a straw from Tonga through the center of the earth it would come out in southern Algeria.

Dennis and Deb sat on a back table, swinging their legs during the lecture. The Welshman noticed them, thought, What an odd couple. A pretty girl. A Jesus-freak guy.

Dennis was high that night, for it seemed that Deb was affectionate toward him. But when he said, "What are you doing tomorrow night?" she said she was busy.

Tuesday night offered protection. Jon Lindborg and Mark Stiffler came over for dinner that night, and Deb made curry for them on her one-burner kerosene stove, and they went over plans to go to New Zealand when school broke in December, and to Ha'apai the following year. Then after dinner she bicycled downtown with them to the movies.

The Finau was showing *The Man Who Would Be King*, the movie based on the Kipling story. Sean Connery and Michael Caine were adventurers going into the wilds of Kafristan, where Sean Connery saw the usefulness of setting himself up as a god unto the people, till a woman betrayed him as a human being. Dennis went to see the movie the next night, Wednesday.

Deb's foreboding about Mary was accurate. On Thursday, she stepped into the corner office at the Peace Corps, and Mary turned down her transfer to Ha'apai. Deb's behavior was not suitable to life in an outer island. "You make a spectacle of yourself, you are too flamboyant," Mary said. Deb would not be respected in an isolated village, she would not be safe.

Deb tried to argue, but when Mary was convinced of something her voice rose, she became cross and brittle. Altogether Deb was behaving badly. Rob Beaver, her supervisor at Tonga High, had said she was struggling as a teacher. If Deb wanted to stay on for a second year, she would need to repeat her cultural training with Tonga 18, the next group of teaching volunteers who were due out in a month's time.

Deb was in a bad way when she came down the stairs, and Frank called out to her from the steps of the mosquito lab. She wheeled her bicycle over to him and cried.

"I'm not suited to be a Peace Corps volunteer. I'm too flamboyant and my conduct is inappropriate to an outer island."

"What? That's crazy," Frank said. "I spent a year in Ha'apai, and you're just as suited as any of the rest of us."

Frank said that things would work out, but Deb shook her head. "I think this is it for me."

"That woman has always had it in for you."

Deb shrugged but didn't disagree, a lot of people had said as much to her.

"Well, I don't know why," she said.

"I do, she's jealous. You're young and real and there is nothing real about her, she's the quintessential D.C. politician with all her makeup. She comes into a room and it's 'Hello, it's so good to see you,' and nobody likes her and right across the room there you are smiling and laughing and there are six or seven people around you laughing too. So of course she's jealous."

Frank wanted to hug her but he realized he'd be taking advantage of the situation. He told her to come along with him and Jackson and Wyler to *The Man Who Would Be King*. Deb had seen the movie already, it was good. Besides, she was busy, she had weaving class.

Weaving class was Thursday night at the University of the South Pacific Centre. The university was based in Suva, but maintained a branch in Nuku'alofa, next to Tonga High School. Several *pālangis* had expressed interest in learning Tongan crafts, and the USP had found an English-speaking woman who could teach weaving.

The course was now several weeks along. Deb was three feet of her way down a mat she was making for her mother. Deb showed her mat off to Talahiva, and Talahiva said, "Deb, you are a better Tongan than I am."

After class, she went out to her bicycle in the USP yard, and there was Francis Lundy. He was dumping trash in the bin near the flagpole. Francis didn't live at USP, his girlfriend Vickie did, and Francis hung out there. They started talking about her troubles, and Francis said, "You need to talk to Vickie." So Deb wheeled her bicycle across the yard to the row of little apartments.

She spoke with Vickie and her neighbor, John Myers, another teacher at Tonga High, then went inside Vickie's place to talk some more, and Francis came in and out of the house as the women talked, and tried to give them privacy.

The house was being watched. There had been a break-in at Vickie's *fale* two nights before. The intruder had removed a number of glass louvers then gone inside and rearranged some of Vickie's things, looking for something or other. Vickie reported the crime to the police, and the police had posted a constable named Naeata to the USP Centre. Naeata had climbed to the top of the water tower on the USP grounds and now lay there surreptitiously.

Then it was nine-thirty, and Deb got on her bicycle. She turned left into the King's Road, turned right past the Copra Board.

It was dark, the moon was rounded. In Ngele'ia the rugby field was gray in moonlight, and the wind moved through the mango trees, stirring the new green mangoes on their bumpy cords. The bicycle rider passed through the sweet narcotic air of the datura, its limp flowers hanging down from the shadows like gloves, and the wind carried the idle sound of a guitar, and a hymn sung by boys.

In the bush across from Deb and Emile's huts were two Tongan families. One occupied a cement-block *pālangi* house, the other a *faletonga* with woven coconut walls.

A shop clerk owned the solid house. Vaka Pasa was an evangelical Christian, and when his pastor had broken with the Assemblies of God congregation in town, Vaka had invited him to build a house alongside him in the bush. So that was the Tongan *fale*. The minister, Pila Naufahu, lived there with his wife Le'ota and their baby girl.

Pila held prayer meeting in the house on Thursday nights. A handful of boys came over, and under the one light bulb on a wire stretched from the Pasa house, they sang hymns and spoke in tongues. Pentecostals believe that living in the life of Christ is not something for Sunday but an everyday relationship, a consuming passion, and the boys had all been born again in Christ, Pila had water-baptized them in the lagoon.

So a good-natured smooth-skinned boy called Havili lost his grin for once and shut his eyes and became stiff in the shoulders, and gave his tongue to the Lord. The Holy Spirit came like a mighty rushing wind, as it says in Acts, and Havili's mind did not understand but God knows. He went for a long time in that state, an hour, more, speaking and rocking with his eyes shut, not feeling afterward that more than three minutes had passed. And as for the joy Havili felt, that cannot be expressed in words.

A couple of others also spoke in tongues, then they sang a hymn, "Sisu Pe Koe Tali." Jesus is Sisu in Tongan—there is no *J* in the King's alphabet—and Pila gave a homily on that theme, Jesus is the answer, then it was nine-thirty and the service wound down, and Pila and Le'ota went outside to the tank alongside the Pasa house to wash their faces and get ready for bed.

Le'ota saw two *pālangis* in the doorway on the other side of the track. Deb was there, or Tepola, as the Tongans pronounced her name, in a nightdress alongside a *pālangi* with a beard whom Le'ota had seen before.

The light fell on his big square head, on his glasses.

Le'ota and Pila went back inside. Only the kerosene lamp was going now. Sekona had cards out to play *swipi*. Sateki picked at the guitar. To'a said good night and went out to the house next door.

To'a Pasa was a tall boy with wideset eyes. He was a second son, self-contained, and seemed older than his 16 years. At night, he generally left Ngele'ia to walk with his cousin Takitoa to Takitoa's house in the old city, so that he could rise near his school, the Anglican high school, and get there fresh in his *ta'ovala* and skirt.

To'a packed his things. He put his blue skirt and books in his bag along with a cartoon book, then went out to the tank at the corner of his house to brush his teeth. He laid the toothbrush back on a high shelf and Deb screamed.

Sounds came from behind her door, a banging, another high shriek, and though To'a was terrified, he ran through the bush to stand at the dirt track. The shrieks came again and again, and To'a made a cry of his own, he cleared his throat, made the sharp glottal rattle that Tongans make when they are telling dogs or children to stop what they are doing, now.

The door opened inwardly and a bearded man came around its edge with a woman in his arms, as if he meant to drag her off.

It was Tepola in a white nightdress, and To'a knew the man, he had seen him often with 'Emili.

The man saw To'a and dropped Deb in the doorway. He took a bicycle from the side of the house and wheeled it off toward the rugby field. He had on only one flip-flop.

Deb lay where he had left her, face down. Her back was bleeding.

Now the others came. Le'ota had thought it was a cat shrieking at first, she had told the boys to be quiet. Then the scream was continuous, and they rushed out. Some boys were afraid to approach. Takitoa stood back behind a *tangitangi* bush. He had seen the *pālangi* man through an open shutter of the house, his arm raised before it came down like the arm of a man in a movie.

Le'ota sat beside Deb and Deb lifted her head. She was badly hurt. There were cuts all around her face, blood flowed from her neck down over the white dress. Her little house made a hideous scene. Sprays of blood had hit the walls and streaked the mats, possessions were thrown about. A dagger with a silver-knobbed handle lay on the floor covered in blood, a pair of eyeglasses was on the mat, broken.

"Please take me to the hospital," Deb said, in Tongan, and in English.

A boy was commanded to run and call the police, and there followed a moment of indecision that would haunt a couple of neighbors the rest of their lives. Should they touch the girl or not? The police would be sure to blame a Tongan boy for this, that was how the police worked. They would beat the living hell out of Tongan boys to learn who had done this. Maybe the neighbors should leave the girl as she was so they might say, This is how we found her.

That was crazy. Still, that hesitation passed among them, before Deb said that they must take her to the hospital, please—*Fakamolemole 'ave ki falemahaki*—and the Pasas' old green truck howled, reversing itself through the bush and boys jumped into the bed to take out the *'ufis*, the long yams, flinging them helter-skelter on the ground.

Le'ota took her shoulder, and to the four boys who also carried her over the low fence, 'Aleki, To'a, Havili, Edgar, the shock was the more intense as her flesh pressed against them through the thin white fabric. How often they had tried to see naked girls. A Tongan boy, even one water-baptized and speaking in tongues with the Holy Spirit, will climb trees to see a girl bathing or half-naked in a window. Now here was a girl almost naked, and that confused them because it did not matter at all, and though Deb's dress fell away from her thighs and more as they lifted her, Havili reached out to cover her, and they laid her down in the truck bed.

A burly man with muttonchop sideburns had come running up, he held Deb's shoulder. The news had flooded out through Ngele'ia, and Pila Mateialona was a Peace Corps driver.

"Tepi where is 'Emili?"

"He wasn't here—"

"Who did this to you?"

"Dennis."

Pila saw blood leaking from a flap in her neck and put his fingers against it. "Be quiet now, we will take you to the hospital—"

"Make sure that they take me home, Pila, will you take me home?"

"Of course, don't talk."

She knew she was dying. Now the truck started from Ngele'ia with Deb in Tongan hands. Le'ota held her head and stroked her bloody forehead and told her she was going to be all right, Havili held her arm, To'a held another arm, 'Aleki sat at her side. Please, God, help Debbie straightaway, 'Aleki said, while To'a said, *Fa'amolemole 'Otua, fa'amolemole 'Eiki.* Please, God, please, Lord, let Debbie have more time on Earth.

Havili began a hymn under his breath, and To'a picked up the thread of the melody.

Maybe say a prayer for the truck, too. It was pushing 30 years old, Vaka Pasa had bought a wreck in New Zealand and shipped it back to farm with, now it was an ambulance bumping fast on potted coral. They passed the mosquito lab on the bypass road and Deb sat up, seemed to reach for something, gasped. *'Osi mate*, Le'ota thought, she is dead. But this was not the case, though she was fading. Her

eyes were wide open and flat, open to the sky and the mess of southern constellations that are nothing like the northern constellations, and the palms that were on
one side of the road but not the other, the lagoon side.

No one can say what she was thinking, anyone can say what she was thinking.
She was thinking about her mother and father, she was thinking about Craig.

The nurses heard the Bedford from far off. Vaiola Hospital is up a hill from
the lagoon, they heard the engine screeching down the King's Road, heard the old
thing swerve right and grind through gears gasping and they gathered under the
portico at the back, the emergency room door. The nurses had been informed by
phone. Sateki had found a phone at the Wesleyan church at the edge of Ngele'ia,
the police had alerted the hospital.

Lutui came running from obstetrics. He had been attending a woman in
labor, now here was a woman being carried across the threshold of death. Later
there was a story that Lutui fainted from the sight of her. This was not true, he
had seen worse in Fiji during training, but Lutui was in shock. Her hair was so
thick and beautiful spilling off the edge of a gurney, but her dress was still white
only around portions of the shoulder and the edges of the wide skirt.

He hunted for a vein, ordered the nurses to cut the dress away and find saline.
"Call Dr. Puloka, call A-Three-Zed."

Still a faint pulse. If he got saline in he might maintain some blood pressure,
and in doing so he put aside the knowledge that this was pointless. Hospitals in
better-off parts of the world stocked a dried plasma product, but Vaiola didn't
even have a blood bank, it lacked the refrigeration needed. That is why the nurses
had called the radio station. When someone needed blood, a call went out on
A3Z for donors to come to the hospital.

They had stopped the bleeding, they had stripped away the dress, now they
waited for Puloka.

Lutui had saline in the left arm, he worked on the right. He touched the girl's
cheeks. Her nakedness and wounds pressed him down into silence, he looked into
her eyes with love. There were jagged pink cuts everywhere, on her breasts, on her
neck, on her forehead, cheeks, arms, sides. Who would want to do this to such a
person?

She was taking breaths but did not seem to be fighting for breath, still Lutui tried to keep her soul from going. You're not handsome, Taniela, but you are deep, a woman had once said that to him in Fiji, now Lutui tried to go as deep as deep could be into the girl's eyes to hold her.

"Dr. Puloka is here—"

A small light-skinned angular man in his fifties appeared, wearing black plastic–framed glasses.

Puloka was the leading surgeon in the country. He came from a professional family, a high family that had served royalty for generations. He had trained in Fiji as all Tongan doctors had, then done a special training in Australia and now performed what surgery Vaiola could support. On the extremities, now and then the abdomen. He set to work, but Puloka could not deceive himself, he knew the girl was lost. The eyes were half open, the whites were dry. He shone a light in the eye and it did not move. Cardiac arrest was imminent. He touched the wounds, and the thought came to him that whoever had done this to her had had knowledge of the body, had meant to bleed her life slowly from pressure points.

As it was Thursday, the *Chronicle* had come out that morning, with tantalizing news. Opening at the Hauhau that night, the open-air theater where it was necessary to run to the back to get under a little roof when it rained, was *Pussycat Pussycat I Love You*. A note beside the ad said, "Pending censor's approval." All films that played in Tonga were censored at the downtown police station. A committee of sober men led by 'Akau'ola, the police minister, met once a week to cut licentious material.

Emile had set out for the Hauhau by himself, then found a date. There was always someone he knew near Yellow Pier. He paid for a girl's ticket and they watched the show, which proved to be a disappointment, before he walked her back across town and stopped at the Tonga Club for a Steinlager, and another.

The talk at the club was Rick Nathanson. He'd gotten a girl at the Dateline dance the previous Saturday, or the girl had gotten him, and only when he'd walked her home to her place beside the Finau and she had dropped to the floor in front of him and undone his belt was he able to turn his flashlight on her face and discover the five o'clock shadow. All this the young reporter had related to volunteers at the club in the belief that he would command their sympathy. But they had laughed hard at him, and the joke had run well into Thursday.

It was almost eleven o'clock before Emile got to the fence at the town end of the rugby field and two boys came running.

"'Emili, 'Emili. Your friend hurt Tepi. Tenisi."

Emile spilled his bicycle and ran. He'd get a towel and hold it to her bruise, he'd comfort her then deal with Dennis. He had the impression of too many Tongans standing in the track in the dark, and he pushed the door in and imagination was blinded.

A voice said Deb had gone to Vaiola. Emile reached in to turn off the light and get her key from the hook, then locked the padlock to the hasp.

He had never ridden so fast. He didn't worry about the potholes or the dogs that took over the town, he felt that he was flying and all the way he bargained with the wind, that there was not really so much blood. The city was shutting down. From house to house he could hear A3Z signing off for the night. It played the national anthem, a stately funereal piece with tubas.

A policeman stopped Emile outside the doors. The windows of the emergency room were painted gray, but the windows in the doorways were clear, and he was able to see into a hallway and get a glimpse of Deb on a gurney, her dress cut away, her shoulders tilted up.

Well, she would not be tilted up if she was about to die.

The nurses stepped away first, having nothing to do. Perhaps also they were more realistic. They picked up wrappers from the floor, placed scissors and a syringe in a pan. Then only Lutui still had his hand on her shoulder as Puloka also stepped back and, lifting his heavy glasses, squeezed the bridge of his nose. 'Osi mate, he said.

10
The Search

The news was Tongan first. It passed from one Tongan to another, then on to the *pālangis*. And from the beginning, Dennis's name was attached to the news.

Losimani Kapukapu was an emergency room nurse. She was married to the top Tongan on the Peace Corps staff, the administrative officer Viliami Kapukapu. Losimani called her husband from the hospital, and Viliami, a wide-framed graceful man with a long sloping Easter Islandish face, drove fast to Mike Basile's house a few blocks west of the palace along the sea and drove right up on to the front lawn alongside the porch, ran up the porch stairs, pounded on the door.

"Mike, Mike, wake up!! Wake up!

"Something very bad has happened."

Mike pulled on his clothes and rushed out without knowing why, and the figures of Viliami and the green Peace Corps pickup on the lawn seemed to him to come out of a dream. Viliami was breathless.

"Debbie Gardner is dead. She was stabbed to death by Dennis."

Mike's mouth was dry as sand. "Where are we going?"

"We have to find him, he may kill himself with his knife."

Another nurse at Vaiola had called Mary and said, "A Peace Corps volunteer has been injured." When Mary got to the emergency room, Lutui lifted the sheet, and Mary gazed at Deb for a time before turning away.

"Where is the phone? We have to find Dennis. He could kill himself."

A nurse pointed out the phone, and Lutui's eyes followed Mary across the room. But inside he said, What is wrong with you? You see how she has been mutilated and you are thinking about the killer? It will make things easier if he kills himself. Good luck to him.

And all the while Lutui's expression was mild, showing nothing of his feelings, in the impassive superior Tongan manner.

Frank and the other Tonga 11s were headed home from *The Man Who Would Be King*. Wyler had a Peace Corps Land Cruiser. He was driving past the Royal Tombs when a car came toward them, flashing its lights and weaving back and forth in the King's Road as if it meant to hit them.

"Goodness, someone's in a hurry," said Wyler.

Then its horn blared, and it pulled over sharply, and Wyler said, "My God, that's Mary George."

The men got out, Mary sat holding the steering wheel and staring at Frank, a fixed, glittering-eyed look.

"Mary George—what is wrong?" said Jackson.

"Debbie Gardner was murdered."

Frank was berserk, he rushed at Mary's car. "What are you talking about?" But Jackson threw his arms around him, six-foot-four Jackson, and Frank said, "Let me go, let me go, let me go—"

Still Jackson kept his arms around him even as Frank twisted toward Mary's window.

"What happened?" Wyler said.

"We don't know much at all. She was brought to the hospital by her Tongan family, she bled to death."

Then Wyler climbed into Mary's car, saying he better drive, and brought her to the Peace Corps office. In the lot outside the mosquito lab, Frank got to Mary.

"Where is Dennis?"

Her eyes were hard. "Why? What do you know?"

"I don't know a thing."

Wyler went on to Mike Braisted's place. Mike and Mark Stiffler lived in one of the volunteer *fales* behind the McMaths'. Wyler woke them up banging on the door, then Mike opened the door, and from his bedroom Mark heard only deathly silence in the front room. He got on his pants and opened his bedroom door. Mike gazed at him.

"Debbie Gardner's been killed," he said quietly. "We're going to the hospital."

"I'm going with you—"

Then they stood out in the hospital lot, trying to get in to see Deb, wondering what drunken Tongan had done this to her.

Frank's statement to Mary that he knew nothing suspicious about Dennis was not true. At 8:45 or so, the karate picture having ended, there was an intermission before the Finau screened *The Man Who Would Be King*, and the three Tonga 11s stepped out for cigarettes. Frank went to the curb to smoke, and Dennis pulled up on his bicycle, and Frank asked Dennis about a book he'd lent him, and he said he'd given it to Deb Gardner, Frank could get it from her.

Dennis seemed in a different world. "Where are you going?" Frank said.

"Where everyone goes eventually—" And something else, about a spirit, and Dennis pushed off on his bicycle headed east, the opposite direction to his house.

Of course Dennis was also suspected in Ngeleʻia. Toʻa and Leʻota had seen him often before, and when they described him to Sateki, Sateki said that it was Dennis. Dennis had tutored Sateki in math at Emile's house. When Sateki saw the broken glasses on the bloody mat, he was sure of the identification, and then Deb had said Dennis's name to Pila Mateialona.

The police dragnet began. Radios carried the description of a bearded *pālangi* of muscular build, unfriendly, considered armed and dangerous, and a Land Rover drove wildly around the USP Centre till Naeata Manu who had been atop the water tower descended to join the hunt.

The police effort was not very focused. The police knew so little about Dennis, it would be hours before they even got to his house. And the opinion that Lutui had formed at the hospital was an opinion that many policemen now shared: Let him kill himself, it will make things easier. Tongans had a folk belief in something called *malaʻia*, a sort of karma, that a man who caused serious injury to another was cursed, and his descendants, too. Their belief in *malaʻia* mingled with their Christian belief in the Ten Commandments. The murderer would not be able to live with himself, he would soon kill himself or, if he did not come forward, was sure to die in a freakish way. He'd be crushed by a piece of farm equipment or broken on the reef and eaten by sharks. Somehow the murderer would die.

Far better that Dennis kill himself than put the police in the position of having to kill him. For anyone who killed someone in the way that Deb had been killed would hang, and the hangman's job would fall to a policeman. No matter that it was the policeman's duty, blood would then be on the policeman's hands. So *mala'ia* to him, too. Far better that Dennis do the job himself tonight and leave the police to find him in the morning.

Americans were looking for Dennis, too.

From the beginning of the case to its end, the Americans carried out a parallel investigation. At times this was an organized and concerted process, at other times it just seemed to happen. Now it began. Even as police Land Rovers motored about Nuku'alofa in an urgently idle manner, the Peace Corps also began to search for Dennis in a more focused way. Two search parties set out.

Viliami and Michael Basile stopped at Dennis's pressboard house in Longolongo and called his name, then went inside. Viliami led the way with a flashlight.

Dennis's bed was neatly made, his short-wave radio was perched on the stud board at its head. The hut was in precise order. Not a crumb to be seen, no dirty dishes, no dirty clothes.

The belief had taken hold among the *pālangis* that Dennis might try to dispatch himself at a southern beach. Not being entirely sure how to get there, the officials drove first to a high school near the beach. Mike Basile was silent. He thought of Dennis's knife at his belt when he had come to his office to recruit his help, Dennis so shy that he generally disappeared. But he had not disappeared. There had been intimations, had there not?

Tonga College was a middling high school, four miles south of the city in the village of 'Atele. Several volunteers lived there, and a mile away through the bush was the volunteers' favorite beach on the south coast.

That night there had been a dance at Tonga College. Afterward a group of volunteers had gone back to Jon Lindborg's *fale*, under a flamboyant tree, to drink some *hopi* and listen to music, and party some more, maybe smoke something. Kay and Dave Johnson were there, the southern Californians.

Mike Basile's taut face at the door stopped the party like a switch.

"Has anyone seen Dennis?"

"No.""Why?"

"We're concerned that he might harm himself. Who knows 'Atele beach?"

Jon volunteered, Jon often went diving at night for lobster. So they went out to the pickup, and only then, as they were bouncing down the gullied dirt road and the headlights jumped this way and that over the boles of coconuts, the huge poised shields of the taro leaves, the thick worn rope attached to a cow—only then did Mike turn in the front seat and give Jon the news, and Jon wept.

He led the older men along a cliff edge. The wind buffeted them and they could feel the great void at their left. Tongatapu was most elevated on its southern, windward coastline. Here and there the sea had broken down the coral shelves, carving out secluded sandy pockets to which the volunteers bicycled on a Sunday, when it was officially illegal to swim. The wind tore and thrashed the palms, and Jon Lindborg wondered what he was doing here.

Debbie Gardner was dead. Deb who he was planning to go to Ha'apai and New Zealand with, Deb who he'd gone to desert islands with, Deb who he had a high school crush on. When had it begun? Maybe here, on the Fourth of July. Deb had made potato salad and chocolate cake for a bunch of them, somehow Mary had gotten them T-bone steaks. They'd walked back drunk through the bush to his *fale* and set off caps with a hammer on the cement floor, listened to Armed Forces radio herald the Bicentennial.

He remembered Dennis coming out here with Deb once, too. Dennis had on a backpack and an extra tire tube strapped to the backpack, afraid something bad would happen outside the city, and Jon thought, What is she doing with that weirdo?

Having given the police Deb's key, Emile started out from the hospital with Pila Mateialona, the Peace Corps driver with muttonchop sideburns, to look for Dennis, and already Dennis's *fale* was a different house from the house Mike Basile and Viliami had visited.

A single flip-flop lay near the threshold.

Emile found a stick in the yard, and they used it to reach inside and scrape the wall, turn on the light. The house was a mess. Clothes and books were strewn about. Pills were scattered on the floor near the bed, also an empty pill bottle, which Emile bent to examine. Darvon. Maybe Dennis was taking that for his stomach. Emile pocketed a couple of pills in case someone would need to know about them. Then he and Pila went on to the Bird Lady's house.

She lived nearby, maybe Dennis had sought refuge with her. But though Emile knocked many times on the door and shouted her name through the wall, the Bird Lady did not stir. That was odd. It was nearly midnight, why was she gone?

This party also made its way to the coast, but to Hufangalupe Beach, a few miles east of 'Atele.

The Land Cruiser bumped out into a dark field, stopped in tussocked grass. The wind was heavy, it pushed the high grass around, lifted it and flattened it. Pila left the headlights on. Before them the field sloped down into nothingness.

They were on the highest part of the island, a cliff whose face the sea had somehow broken through. Then leaving the cliff edge intact, the sea had continued on to crater the land behind it. Emile walked to the brittle sloping edge but was afraid to go too close. The sea was trapped deep down inside, sucking and hammering and shouting. Emile sent his own cry down into the crater but it was swallowed up.

He started to walk around the crater, then stopped. Its outer edge was a coral escarpment arched over the howling crater like a bone lodged in the monster's throat.

It was one of Dennis's favorite places, and it seemed to Emile that Dennis might kill himself from the escarpment.

He called out Dennis's name, but that made no sense. Dennis's bicycle was nowhere to be seen. The beach at Hufangalupe was a couple of hundred yards to the west, now Emile nerved himself to go down to it. The trail was steep and treacherous. When he came down to the sand, shadows flew about him. The wind sent sprays of water and sand across the beam of his torch, darkness seemed to absorb the light.

Emile and Dennis had been so close!

When they first came out, all they did was laugh together about this place. Dennis's laugh was infectious and sincere, Emile had never heard him fake-laugh. Dennis had taken pleasure in so many things in this out-of-the-world place. One night they went to see a Bronson picture at the Finau. The rows were too close together, Emile's knees were jammed against his chest, which amused Dennis no end. A bare light bulb hung down on a wire from the ceiling, and when the movie was about to start, the projectionist danced the curtain back and forth to the rhythm of the music.

Of course, that had struck them both as so hokey, then the light went off and the movie started and it was not *Death Wish* as promised but a spaghetti western, and they laughed at the blunder, till abruptly the film overheated and browned and melted before their eyes, broke off, the screen blazed white, and Dennis started laughing so hard, making Emile laugh, that Emile felt something in his throat, a tickle, and just like that he was sick on the neighboring seats.

"I'm sorry, I vomited on double-C3 and 4," he said in humiliation to the girl at the ticket booth, and she said, "Oh, I'm sorry, let me get you two other seats."

They laughed about that all the way home.

"I'm sorry, let me get you two other seats."

Dennis walked Emile home. Emile had not been able to afford a bike yet, and Dennis walked his own bike as Emile dragged alongside, still sick. The laughs went from one to the other, and Emile said good night and stripped his foul clothes at the door and went inside.

In the morning when he opened the door, there was a plate of fruit on the step and his clothes were washed and hanging on the line. His Tongan family had done that for him without a word.

Is that when it happened?

Emile started down a different path, and all the complaining and laughing he and Dennis had done about the hypocrisy, the Christianity, the backwardness, the silliness, they dropped away. These people have lived in this place thousands of years, they know what they're doing. It was a classic moment in Peace Corps service, the volunteer accepted his situation. Dennis never did.

"Dennis! Dennis! Dennis!"

But only the wind answered, coming from Antarctica uninterrupted.

The news had now shaken a great number of volunteers from their *fales* with knocks, cries, and sobs, and they gathered at the hospital and the mosquito lab. Of course they were in deep shock. They knew so little, and they thought about themselves, they feared that whoever had stabbed Deb Gardner could stab them too, and they hung and clung, and at the mosquito lab the Tonga 11s looked at the collection of knives and spearguns and thought, We can deal with this.

Mike Basile went into the Peace Corps office at twelve-thirty, found Mary standing at the clacking vibrating telex, sending a cable to Washington.

IMMEDIATE LIMITED OFFICIAL USE. LIMIT DISTRIBUTION.
PCV DEBORAH ANN GARDNER DIED OCTOBER 14 1976 2300. CAUSE
OF DEATH BELIEVED TO BE MULTIPLE STABBLED [SIC] WOUNDS.
CIRCUMSTANCES OF DEATH ARE BEING INVESTIGATED BY POLICE.
NEIGHBOURS WERE WITNESS. POLICE TAKING NAMES AND
NATURE OF THEIR INVOLVEMENTS ...

That was disingenuous. No one suspected the Tongan neighbors. But then who could say what was going to happen before the night was through. And Mary had done all that time as an executive secretary, she had softened bad news before. It would help no one to seed a hundred phantom fears in Washington.

Mary wanted to call Deb's parents, Mike was against that. "No, no, no, let Washington handle that—" They should consult the Peace Corps manual, the manual would have something to say. They went to Mary's office, to the two massive yellow ring-binders containing the manual.

No, they were not to call Debbie's parents. Headquarters would take that responsibility, NANEAP and the Office of Special Services.

So they went on to the hospital, and all the rancor that had built up between the two officials vanished. Mike had to support Mary, there was no doubt about that in his mind. They were a team and outside Vaiola's front door was a host of *pālangi* faces looking to them for leadership.

Mary stopped to talk to every volunteer. She held them or hugged them, she brushed the hair out of their eyes, and for the first time Michael admired Mary. She brought the most intense feelings to everyday encounters. Well, here was a genuine crisis and now crisis became her, she knew how to conduct herself, with gravity and poise.

Vickie Redpath stumbled forward, ghostly white and shaking. A thin scream emerged from between her sobs.

"I was just with her—I was just with her—we were talking, just two hours ago."

It almost appeared that Vickie was on the verge of a breakdown, and Mary held her, comforting her.

Police in their gray uniforms and black belts guarded the front door, but Viliami brought Mike and Mary back through the hospital. A square and faintly modernist three-story building, it had been built with the aid of the British five years before. The two officials came to a back room, and Puloka rose gravely from a chair. There was a bed and a sheet, a human form.

It was important to Puloka to show them the wounds. He had made a preliminary examination of the body and wished to explain to the responsible officials what had befallen Deborah Gardner. It was twenty-one wounds, or twenty-two, the numbers blurred.

Many cuts were to her face. She was very pale. Puloka touched the wounds with his finger, probing them a little. On her right cheek a cut went to her eyebrow. Another at the top of her forehead. In the left jaw. In the right jaw. In the cheek below the mouth, but going through to the mouth. Two or three to the neck. Deep wounds alongside the jugular vein and carotid artery. Either of them could have killed her.

"It would not appear that she was raped—" He and Lutui had done a swab.

Now Puloka drew the sheet back to Deb's waist and lifted her arms to point out the cuts on the breasts. Her body had been cleaned, and some of the cuts, however deep, could not be seen at first. They were like slits the lips of which had sealed.

"This wound was the worst . . ."

Puloka rocked Deb's left side up to reveal a deep gash in the rib.

"The up-and-down motion of the truck bringing her here evidently mimicked the respiratory functions, and so she seemed to breathe. It was only upon internal close examination that I saw the aorta pierced."

Puloka had worked in Fiji and Australia. She would have died anywhere. She would have died at any hospital in the world.

The phone rang, a nurse said Puloka's name. The surgeon lifted the sheet back over Deb's shoulders and left the table, spoke in Tongan for a few seconds only before replacing the receiver and regarding the *pālangis* with a formal expression.

"He has turned himself in, Priven. He is at the central police station."

11

Tapu

At noon that day, October 14, he had stepped from the chem lab into the staff room at Tupou High, and the Bird Lady started to say something, then stopped herself. He had the same look she'd seen some months before when he'd been depressed, a gone look. He was pale, he was looking away fixedly, and she thought, This isn't a good day, so she didn't say anything to him, just hello.

His classes ended early Thursdays, and he rode home to Longolongo and gathered his washing as he did every Thursday and brought it to the McMaths. He wheeled the washing machine across the room on its casters and hooked it up to the sink. Then he showed Marie a new shirt. He was worried that the color was going to run.

"When the soapy water comes out of the washer into the sink, hand-wash it," Marie said. "Squeeze it out, hang it on the line."

Marie was not herself that day. She had bronchitis, she was on antibiotics and needed rest. Her youngest, Dwayne, was also home and ill, asleep.

"Now, Dennis, I am going to sleep. If you like, hang your stuff outside and wait for John, when he comes home to tea."

"I think I will wait for John."

Marie went into the bedroom, and Brutus came too, the McMaths' dog with a fishhook tail, and as he always did, Brutus lay down under the bed. Marie slept heavily.

Some time later Dennis left the house and Mark Stiffler, who lived in that little *fale* behind, saw him, asked Dennis if he wanted to play bridge later. But Dennis's eyes looked right through Mark, and he let the door close.

The door woke Marie. Out the window she saw Dennis going down the lane, his washing over his shoulder in a bag. He'd said that he was going to hang up his washing and wait for John. Generally when he did that he read a book, or talked to Marie, or did his marketing.

But he had not done as he'd said. The wet washing was in his bag on his back. That was Marie's first sign that something was wrong.

He could not wish to be arrested. Peace Corps volunteers had heard the same stories that the boys in Ngele'ia heard. Suspects were handcuffed to a flagpole for the night and most of the next morning, sitting outside in plain sight while the police figured out what to do with them. And that was for petty crime. A rape suspect was stomped in the back of a police Land Rover, then tossed into a cell with a basin handed after, so that he could piss blood. When someone stole Emile's radio, a police friend told Emile that he would look for the man.

"If I find him I will severely arrest him," the policeman promised, and Dennis and Emile had cracked up over that.

Midnight found him bicycling along the northern foreshore. The wind was a breeze here, it had beat itself out against the southern beaches where the search parties had gone looking for him.

Now he came to the faux-Tongan *fale* that had stopped the Ambassador a week before, the *fale* sheathed in *kaho* from which Selden had emerged not believing Americans could live there. The place was the most civilized household in Nuku'alofa, as civilized a household as could be found anywhere in the world. Paul Boucher lived here with his wife Laura. They had bridge games, and Scrabble games. The Bouchers played show tunes and jazz on their tape player. Paul taught alongside Dennis at Tupou High and could match Dennis's intelligence. He wore glasses and always thought before he spoke. A mature, detached, analytical person, he seemed to watch even as he participated and didn't seem to have assimilated much in Polynesia. He didn't go to church, and hadn't gone very far with the language.

Paul was up late grading papers, when he heard a sound and went to the door. Dennis was straddling his bicycle. He was not wearing his glasses.

"Dennis, what's the matter?"

Dennis shook his head. "Something bad has happened. We have to go to the police."

Laura was in the loft sleeping. Paul shut the light and locked the house, then they set out together on bicycles along the foreshore.

The Vuna Road was lovely at night. Its coral bed was groomed to serve the affluent who lived west of the palace, it glided under handsome trees with the sea always at hand. A light or two glinted on the slumbering water, coconuts tufted the horizon, islands on the reef. Now the two cyclists made a great bend around the palace grounds past the shining perfect forms of the royal palms in His Majesty's backyard, with His Majesty's white geese roosted underneath.

On they went past the Stone House, where the King's brother, the prime minister, worked, then bumped over a field and under the shelter of a great banyan tree to a little bandstand that stood opposite the door of the central police station. It occurred to Paul that Dennis had caught somebody breaking into his *fale*. Break-ins were almost part of the culture here. Paul was sure that Dennis had gotten into it with the thief, maybe hurt him.

Dennis dropped his bike against the bandstand and reached into his jeans. "Here is my wallet. There's $40 in it, I won't be needing that."

"Dennis, what is wrong?"

"You can have my short wave radio and Paul can have my tape recorder."

He meant the other Paul from Tonga 16, Zenker, the curly-haired scientist from Notre Dame.

The door of the station was open, the front room was milling with officers, and one of them, Naeata Manu, who had been atop the water tank, noticed the two *pālangis* by the bandstand and walked out toward them.

"I don't know what's going on," Paul said to him, but Naeata moved past him.

It took some time to convince Dennis to go inside. He was disturbed, resistant, strange, and on the rickety steps in the light of the doorway it was evident that there was blood on his clothing.

The police station was sagging and antiquated. Its clapboard had been eaten away by termites, its Victorian gingerbread had been left unpainted for too long. The police complained about the station all the time, and the police minister, the fearsome 'Akau'ola, frequently pressed his fellow Cabinet members for funds to replace it, calling it "a disgrace to the administration of criminal justice or indeed to any other administration." The minister's offices were at the other end of town, but he came here often to a dilapidated room where pieces of fiberboard fell from the ceiling, to chair the Cinematograph Censorship Board and ensure that extended families could watch a film together with no infraction of *tapu*. Kisses, not long kisses, might be kept. A thigh was tolerable, now and then a buttock. But never a breast.

The police brought Dennis through a large front room, and around the counter to an official area called the charge room. Paul came too, refusing to leave his friend, he wanted to know what was going on. The police removed Dennis's belt and told him that he must enter a cell.

Dennis sat on the floor and refused to move. It was dark in the cell, he said. He didn't want to go into the dark.

"I want to stay in the light."

"I will go in there with him," Paul said.

"No, you can't go with him."

At last the police made Paul go away, but Paul stayed in the public area on the other side of the counter, sitting on a bench to keep an eye on things. While inside the charge room, Dennis was surrounded by several big Tongan policemen with shiny black belts and red details on their blouses.

A small officer, ramrod straight with a chevron mustache, addressed him. "Why did you come here?"

"I tried to kill myself, I swallowed pills at eleven o'clock."

He seemed to grow fainter by the moment, he was obviously in a state of profound despair. But the officer had been inside Deb's *fale*, and had no kindness for Dennis, so the police dumped Dennis inside a bare cell, cell number 1, to await the arrival of the chief inspector for the entire Kingdom.

Before long he arrived, a big man with a low forehead and very dark skin. The

only thing even faintly soft about Fakaʻilo Penitani were the long glossy curls of
hair on the front of his head, to which he seemed to pay attention.

The prisoner was brought back out from cell 1 and compelled to sit at a long
table stained with ink and sweat.

"I have cut my wrists—"

Had he gotten his hands on some blade in the cell? The policemen dragged
his arms out on the table and peered at the wounds.

"He didn't go very deep," said Naeata.

"No, if you really want to kill yourself, you must cut at an angle," said the
small stiff officer, whose name was Kuli Taulahi. "I don't know that he has even
used a knife."

The chief inspector insisted that Dennis stand up for the examination. He
was deathly, wobbling, and when Fakaʻilo asked why he had come to the police, he
said again in the thinnest voice, "I took an overdose of pills, I cut my hands."

"Why?" Fakaʻilo said.

Dennis lowered his head, his whole body was shaking. It appeared that he
was about to die right in front of Fakaʻilo, but the police were to wait an hour or
more before bringing him to the hospital. Fakaʻilo held up a sheet, read Dennis the
caution.

"You are under arrest. The criminal offense is murder. You need not say anything, but anything you do say will be taken down and may be given in evidence
against you. Do you have anything to say?"

Dennis motioned for a pencil, and was permitted to fall into a chair. Taking
the charge sheet from Fakaʻilo, he wrote, "I have nothing to say."

This he was asked to sign, and did.

They still hoped that they could trap him. Kuli laid out on the table several
items that he had collected from Deb's *fale* and labeling the pieces of evidence one by
one, giving each a letter, placed the exhibits inside a cardboard box. Broken eyeglasses. A black-handled serrated knife called Seahorse. A single flip-flop. And
other things as well that the boys had not seen as they stood at Deb's doorway, afraid
to enter: a syringe, a metal pipe, a bottle containing powder, a jar, and inside that jar
a glass bottle.

Kuli started to twist the top off the jar to get at the bottle and Dennis tilted sharply forward in the chair.

"That is *tapu*—"

"*Tapu?*" Kuli could not be sure whether Dennis was joking.

"*Tapu. Tapu.* That can kill hundreds of people."

Kuli set the jar carefully down in the box, and Dennis was dumped back into cell number 1, when there was a flurry in the public area. Mary had arrived. She was able to see the prisoner for all of a moment. Oh, how Dennis loathed Mary. Now he had a request of her. "I need a psychiatrist."

Then who should come next but Emile. He was allowed to look in on his old friend. Dennis seemed half his normal size. He wore a faded orange Illinois T-shirt and his jeans cut-offs. Blood was smeared on the leg.

"Dennis—I have just one question. Why?"

Dennis didn't look up. "I'm insane, and this just proves it."

Did anyone sleep that night? A vigil had begun on the steps of the mosquito lab. Men passed in and out of Jackson's apartment, crying, or sat stunned on the steps, staring at the gray-green lagoon.

Mike Braisted and Frank Bevacqua were as close as they had ever been. They walked around the city together looking for friends, till Frank said that he needed to be by himself and he sat out on Yellow Pier and when the wind came he heard Deb's wild laugh rising from the sea to seize his spirit.

Emile and Paul Boucher left the police station together, and Emile said, "Did what I think happened really happen?" and Paul nodded his head, "I think so." But Paul was Dennis's good friend, Emile had lost that position months before, and later that night, Mike Basile and Mary George went with Paul to Dennis's house, to see what clues they might find to Dennis's state of mind. The police had not even been there.

After that Mike embraced Mary and said good night. He would see her in a couple of hours, they would make the rounds of Tongan officials.

Pale dawn was at hand, and the feeling existed throughout Nuku'alofa that other horrors of the night were yet to be discovered. No one had been able to find the Bird

Lady. She was not at home, had never returned. Had something befallen her as well?

The McMaths stirred at six or so in the morning, and three or four of their volunteer neighbors called out to them from the fence, and they came outside.

Whereupon Marie buried her head in her husband's neck, crying, and John said that he was going off to see his teachers. "Lock the door and keep the boys and Brutus inside," John said to his wife, and then he set out for Gay Roberts's *fale*.

Gay was the New Zealand volunteer for whom Dennis had made the mailbox key, she lived near the palace. But another teacher from Tupou High School beat John there. Matthew Abel crashed his bicycle outside Gay's door and burst into her kitchen at 7 A.M., threw himself into a chair.

"Dennis murdered Debbie Gardner last night—"

Then gathering his breath he looked around at the three women having tea, Gay, Gay's roommate, and the Bird Lady, alive as can be. She'd spent the night at Gay's. The Water Board had cut off her water the day before and after calling the board in anger, she'd closed up her house and left.

Now came John McMath, white as a sheet, to repeat the news, and no one said anything at all for a time, and the Bird Lady who was ordinarily so thoughtful was stunned into blankness. Words would not go into her brain, and it seemed to her later that it was like the Kennedy assassination. Life had been cleaved in two parts, the time that Dennis was incapable of murder and the time that Dennis was said to be a murderer.

The last to be informed were the new volunteers, Tonga 17. They were living at Sela's, they were still in language training. They had been here only twenty days and had not stitched themselves into the existing social web, no one had awakened them during the night.

Their language class was upstairs in the Kapukapu house, the Peace Corps administrator's, a large room with louvered glass windows on all sides and cushions on the floor to sit upon. Mary arrived at nine o'clock, looking haggard and noble, her face blotchy.

"Last night we experienced a terrible tragedy that will have implications for all of us in Peace Corps in Tonga, and quite possibly implications for the Peace Corps worldwide. A volunteer died of stab wounds, Debbie Gardner—"

Shouts, cries. Silence.

"And another volunteer has been charged with murder."

The name Dennis Priven was offered, but though a couple of people had questions, Mary said she didn't know very much at all. Sitting on a cushion, Caroline began to sob, and Rick Nathanson tried to console her. "What, Caroline, did you know her well?" Caroline shook her head, covered her face with her hands.

Mary had to get to the office, she had important calls to make to headquarters. It was after nine in the morning Nuku'alofa time, Friday, October 15. That made it four in the afternoon on Thursday in Washington. She needed to determine what Dennis's rights were, and find him a psychiatrist, and what with the radio connection, the calls were difficult. On her first try Mary ended up at the University of Maryland.

Between calls, Mary told the Tongan staff and a volunteer about seeing Deb's body at the hospital.

"The look on her face—" Mary shut her eyes. "A look was frozen there. It was horrifying."

A police Land Rover came into the lot. Faka'ilo got out and climbed the staircase. The chief inspector was en route to Vaiola Hospital. The autopsy was about to begin. It was the law that a police officer be present and that in the officer's presence, someone who knew the girl identify her body. Mary George was the head of the girl's organization, she must perform that role.

A constable sat waiting at the wheel of the Land Rover, tried not to look at the tear-swollen faces of the volunteers strung up and down the mosquito lab steps. It was only two or three minutes before Faka'ilo came back down the steps alone and slung himself into the seat and gave a nod. He seemed furious.

Faiva eased the truck into the Bypass Road.

"That is the very first time I have heard anything like this," Faka'ilo said. "The Peace Corps directoress has refused to come."

"Refused?"

"Refused. She said Peace Corps is concerned with the defense of the one still living."

12

"Help Me Tell the Untold Story
of the Murder of Deborah Gardner"

The plane left Samoa on a perfectly clear night in March 2001 and as it started to descend I made out lights on little Lifuka, the island shaped like a shark's tooth that Deb Gardner had wanted to transfer to, then we went farther down and there was the city so many former volunteers had described, bands of lacy light strung along a slender waist of land between the black sea and the black lagoon. Such a small city to have made so many memories.

The King was still the King. At 7:45 the next morning, I went to the palace office to drop off a letter informing His Majesty of my project. A sweeper at the Stone House gave me directions to the attorney general's office. I was there by 8, waiting for him to get to work. I was told to come back at 10, and then Tevita Tupou told me about stabbing the Seahorse knife into the prosecution table as he summarized his case a quarter century before.

Time moved in a different way in the tiny country. The past was not sealed off. Everywhere I went, in every *fale* and bar, I met people who still thought about Deb Gardner, or brought me to someone who did. "You are raking up my heartache and boiling the anger in me," said 'Asinate Matangi, the former sixth-form prefect at Tonga High. "She was just Debbie, and she was pure, and I am actually shocked—I can feel her presence, it is like she is trying to—oh, we wanted her revenge, but somehow it never came. I was a baby, and then I was toughened forever."

After that first trip it was hard to stay away. On my second visit, in June 2001, I published a letter in the *Chronicle* asking Tongans who had information about the case to call me at Sela's Guest House. The matter had left serious questions of international justice unanswered, I said, and my country was a great democracy but the story had never been told there. It was the first time in twenty-four years that Deb and Dennis's names had been published, and though I worried that the *Chronicle* might somehow make it back to the United States and to Dennis, it seemed worth the risk. Paua, Rick's old editor, wrote the headline: WOULD YOU HELP ME TELL THE UNTOLD STORY OF THE MURDER OF DEBORAH GARDNER? and published the letter in English and Tongan.

It was like the front page of the *New York Times*. Everyone read it, even the King's daughter, Princess Pilolevu, and people called and wrote, including two boys who had been in Ngele'ia that night. One boy led to another. They were all ministers now, or just about.

I showed them a picture that Emile had given me. It was like a lineup from volunteer wonderland, five *pālangis* laughing in the bush in October 1975. A sunflower teetered over Emile's house, four young banana trees sprouted from his lawn. Lampshade Phil had a flower in his teeth, Emile wore a white skirt and held a bush knife, John Myers wore sunglasses, and John Sheehan wore a flowered red skirt and gave Dennis rabbit ears. Dennis stood at the center, smiling softly, holding Emile's guitar by the neck.

Deb would not show up in the Kingdom for another month, but Dennis had on the bright orange Illinois T-shirt that, greatly faded, would be stripped from him by police a year later.

None of the boys had any trouble pointing out who they'd seen.

"It is a moon night," To'a Pasa said. "You can see everything, it is clear."

To'a was now tall and burly, a minister in the Tongan section on the north shore in Auckland, New Zealand. He came to his door Monday morning in a blue-and-white lavalava, and as we talked smoked one cigarette after another down to the filter and stubbed them out in an ashtray on the floor.

He'd stayed for hours at Vaiola that night. "We pray for Debbie's family, we ask for the peace of the parents."

Lutui lived across Auckland, in Morningside. He wanted Deb's photograph. I gave him one that Frank had taken, Deb outside and wearing an oversized denim shirt, her thick hair all over her shoulders.

"You know, I treated him that night as well. It was my duty. He tried to kill himself but he found it was too painful. He was a coward. Then he went back to the States and was killed."

"No."

Lutui tilted his head in surprise. "We were told that he was dead."

"That's not true."

"But as soon as he got off the plane, he was killed. Someone shot him. Maybe someone hired by the girl's family."

"No, he is free. He lives in my city."

Lutui was quiet.

"Well, I can never forgive him. Because of what I saw that night."

"You said you were a Christian."

"If he submits himself, I might forgive him. If he apologizes to her parents."

I drove back into the central city and went to an Internet place, found a note from my wife. *Someone called you might be interested in—Wayne Gardner, Debbie's father. He wants to help.*

13

Next of Kin

At 6 P.M. on October 14, 1976, John Dellenback stepped from his office and in a soft voice asked several people in the outer office to join him inside.

"I'm about to do one of the hardest things I've ever done in my life. I think it would be good for all of us if we gathered together."

Three or four assistants and secretaries now joined him, and glanced out the window at the White House. From up here it looked like a big flat wedding cake, the long lawn stretched out behind. They took chairs, a couple arranged themselves on the couch, and a secretary, not wanting to look her boss in the eye at such a stressful time, focused on the picture of Jesus on the wall. It was a portrait of the savior with golden hair and a red robe, similar to portraits on the walls of Tongan *fales*.

"Is Alice Gardner there?"

"Well, would you mind getting her? And please, do me a favor—will you stand by when I talk with her? I'm about to give her desperately sad news."

Headquarters was in a tall gray building on the north side of Lafayette Park. The closeness to the White House had always conveyed a sense of privilege to Peace Corps workers. From time to time, the White House invited them to the Rose Garden to fill out receptions for foreign leaders, and in Peace Corps mythology, the first Peace Corps director had walked across the park to attend Cabinet meetings with his brother-in-law.

Of course that was long ago. The New Frontier was gone, a sacred relic. The Peace Corps director was not even called the Peace Corps director anymore. John Dellenback was Associate Director of Action for International Operations, and reported to the Action director, a man with a crew cut on the fifth floor. The staff bitterly resented the bureaucracy in which they were immured. How many Americans had heard of the programs Peace Corps was stuck with, Foster Grandparents, or even Vista? But in addition to bitterness there was fear. Peace Corps was being left to waste away and die an anonymous death, and the staff were like true believers under long siege.

Their leader was suited to that role. A wealthy and religious man who favored bow ties, John Dellenback had led prayer breakfasts on the Hill until the Watergate election in 1974 had swept even moderate Republicans like him from Congress. Then he did not go back to Oregon, but hung around the capital. He was one of those decent nonpartisan men that the political world likes to have around if only to help redeem the image of politics, and after Watergate Dellenback conducted prayer breakfasts to help Washington.

The new President's friends leaned on Jerry Ford to give Dellenback a job, and in the spring of 1975 Ford came to a prayer breakfast, and John Dellenback went so far as to drop him a note.

"As John—a struggling Christian, deeply troubled about the state of God's world—I write to you as Jerry—a fellow pilgrim—and say thanks."

So a month later, Dellenback became head of the Peace Corps, and his swearing-in in the Oval Office was more an ordination than an appointment. As if God had called John Dellenback in his late fifties to an important mission that he would take on with love and fervor: saving Peace Corps. He went back to Congress, beseeched his old friends for funds.

"I think, Mr. Chairman, rather than my adding anything—other than this comment, on my strong personal pleasure in being here, my strong personal pleasure in being part of what I think is as fine an expression of the basic American spirit that exists anywhere in this government . . ."

If John Dellenback was pious, he was also political. Even as he fought for Peace Corps, he did not forget who had given him the job, and in the fall of 1976, when

Jerry Ford was trailing in the polls, Dellenback sent a memo to his boss, the young bristle-headed Action director on the fifth floor, seeking permission to campaign.

"Things are apparently heating up with the President Ford Committee to the point where I have been asked to make several appearances on the President's behalf. They include a tour of a good many medium-sized Oregon cities beginning October 20 and running through October 23, and a speech at the University of Connecticut on October 15.

"Unless you feel I should not do so, I stand ready to fill these requests."

The okay came back the following day, October 14. By then the director had other things on his mind.

Because Jack Andrews, the regional director, was still traveling out in NANEAP, his responsibility to call the next of kin had passed to John Dellenback, and the Peace Corps director had waited about as long as he could, six o'clock, then summoned company. His call went into the switchboard at the Bon Marché, a solid red-brick department store that served the affluent west end of Tacoma and the better suburbs, and though the switchboard operator was not in a position to bodily do what Dellenback asked, to stand with Alice as she took the phone, she stayed on the line.

Alice worked in Coats near the elevator, just the other side of Moderate Dresses. She was a formal and cool person, more formal than her daughter. Fine-boned and pretty at 45, with rich dark chestnut hair done at the hairdresser's in the latest fashion, she was seen by the girls of the second floor as the best. She knew her business, she knew her stock, she was fast, she was courteous, she made sure that every button was buttoned.

Women's clothing took up only one part of the vast second floor, and Alice's scream reached its farthest corners. Scores of shoppers turned to see a pretty woman collapse sobbing at a counter.

The switchboard operator now understood the matter, and paged the Floor Divisional, who went running to Alice. As did Justine Corby, Alice's close friend, who worked in Designer Sportswear. Alice stared right through Justine. "Debbie has been stabbed to death—"

Justine grabbed Alice and brought her to the basement lunchroom, got her water, comforted her, then drove her home to the condominium on the cul-de-sac in Steilacoom, and when they got there Justine called a number that John Dellenback had given Alice for the Office of Special Services, which handled volunteer emergencies.

Justine was enraged. She said that what the Peace Corps director had done to Alice was cruel, giving her the news as she was standing on the floor.

"At least get her off the floor! They could have sent someone to bring her into an office. A manager—somebody in personnel."

The woman at Special Services said Peace Corps had no choice but to call the next of kin at work. The last thing Peace Corps wanted was for someone to be driving home and hear the news on the radio, they might have a crash and hurt themselves.

"But Debbie died in Tonga," Justine said. "There is nothing on the radio."

Justine was correct. At this point, few people knew anything, even people inside Peace Corps headquarters.

Mary's first cable had come in that morning, in time to greet Peace Corps workers as they got to work, and though it offered the hope that a Tongan would be arrested for the murder, that prayer had disappeared as the day wore on.

Mary's next cable that day said,

EYES ONLY LIMITED OFFICIAL USE LIMITED DISTRIBUTION
PCV DENNIS PRIVEN SUSPECTED OF MURDER OF PCV DEBORAH
ANN GARDNER . . .

Mary's apprehensions about the news were shared by officials at headquarters. The chief counsel of Action issued a stiff memorandum about the case: "I consider that it is vitally important that all communications outside of ACTION, and particularly to the families of PCV Gardner and PCV Priven be carefully controlled."

Nothing like this had happened in ten years at Peace Corps, and the case was potentially explosive in any number of ways. John Dellenback was about to go out campaigning. It wasn't the kind of thing Peace Corps wanted to tell politicians who were approaching a general election, it wasn't the kind of thing Peace Corps wanted to tell parents whose kids were pushing to go overseas.

The counsel wasn't just saying, Don't talk to the public and the press about Tonga. He was saying, Watch who you tell in government. In fact, one of those who was to be kept in the dark was the Action director on the fifth floor.

Peace Corps church despised Michael Balzano. He was 40 years old, a former New Haven garbageman who had pulled himself up by the bootstraps and earned a Ph.D. at Georgetown. He wore his hair in that crew cut, and he did not like to travel, though when forced to do so, he carried cans of tuna fish with him so that he wouldn't get sick from local food. Fearful for himself, Balzano was also passionate about risks to volunteers—wars, malaria outbreaks, and so forth. So Peace Corps, whose staff was loaded with former volunteers who understood the risks, felt that Balzano did not get the mission. He seemed narrow minded and suspicious, and Peace Corps kept things from him.

John Dellenback was about to hit the campaign trail. First he sent a three-page memorandum to Balzano about everything that had happened that week in Peace Corps, and all but lied about Tonga.

"Significant Activities

"NANEAP. The Regional Director is modifying his travel plans to go immediately to Tonga to assist the Country Director and to determine the need for any additional assistance required in the recent death of one volunteer and imprisonment of another."

Modifying, assist, death, imprisonment. Just don't say murder.

Three other girls from the Bon came over, Ellen Ketcham, Alta Clabaugh, and Alice Gappa, and took Alice out to a restaurant in Lakewood. They had coffee and Alice cried and the girls held her and tried to comfort her. But what was there to say? Alice had been as close as a sister to her daughter, and the women saw Alice stepping through a doorway to the underworld.

In her composed moments, Alice told the girls about Deb's sense of commitment. As a teenager, she had volunteered at Tacoma General Hospital, and every two weeks for more than a year Alice had driven her downtown. She had gone off to Tonga because she followed the beat of her own drum and because she wanted her life to count. Alice's mother Marie had given Deb a small leatherbound New Testament to bring out with her, Alice had given her her gold cross.

Deb wrote poems and had sent some home. Alice read one and started to cry again.

Of course, the girls asked about Deb's father. Alice didn't know much about Wayne now other than that he was in Anchorage. She hadn't seen him for years, not since one of Craig's track meets. One day at Lakes High School she'd been there and Wayne had been there, too, and so was Deb, who was on the drill team then.

It was hard to imagine what had kept them together aside from being conservative Republicans. Alice was elegant, proper. Her parents were religious Swedish Americans, highly reserved people, and Alice shared that reserve. Her mother had worn gloves to the grocery store and held her daughter to the same standards. A roast or a chicken on Sunday, never hamburgers on the grill.

Alice had fallen away from that, of course, after she'd married Wayne. But she hadn't worked while they were married. She respected her husband, and they were strict parents. When Wayne said that the Beatles letting their hair grow out was disrupting family values all across America, Alice held the line as long as she could. She made sure Craig had a crew cut, and once she made Debbie cut her long hair. Deb shoved the permanent under the sink to wash the stinky preparation out, and cried in her room at the loss, and after that, her parents left her hair alone. And Craig's hair? Craig looked like a hippie in high school.

Alice's mother hadn't been able to say the word divorce. She spoke about this one or that one getting a divorce with lament in her voice, till Alice finally said, "Mother, excuse me, I'm divorced. You know that."

She was a single mother, and if she struggled to keep up payments on the condo that she had bought for $17,000, getting a loan for the first time in her life, or to send packages to her daughter in the Peace Corps in the South Pacific, she didn't let on about it. Wayne Gardner was leading a big western life, but Alice was a dignified person who cared about appearances.

Not that she said a word against Wayne. The last thing she wanted was to embitter the children against their father.

Still, she had a sharp tongue. "You preferred an elk to me," she had said to Wayne when they divorced, and he had had to agree, there was truth in that.

As she was in a cold war with her father, when the Peace Corps asked the volunteers to write out their next of kin on a sheet of paper in San Francisco, Deb had written 1, Alice Gardner, and 2, Howard Gardner. Howard was Wayne's older brother, a metallurgist in Richland, Washington, and on October 14, the Office of Special Services called Uncle Howard, who said that his brother was hunting but he would find him and he called Wayne's second wife, Barbara, who answered the call in the kitchen of her suburban Anchorage trilevel.

"Oh my god, are you sure?"

Wayne had left town a day before. October was moose season; in October, Wayne went up to the Arctic Circle to get a moose. This year he and a friend called Warren Fortier had flown to the village of Galena, then Harold's Air Service took them north over the tundra following the Koyukuk River, a tributary of the Yukon.

The river was shrunken, sleeved by broad flat sweeps of yellow gray sand and gravel. The plane kicked in onto a bar, and the men heaved out several packs and coolers, and the plane was gone. The tent was big enough to stand up in. Wayne didn't believe in roughing it. He'd camp for five days or so and hunt moose and live well.

They drank scotch from plastic bottles, Wayne made noodles and venison for dinner. They could hear the wolves howling as they ate, and Warren bitched about his divorce. He felt cursed. He was something of a hard-luck case. He'd moved up to Alaska from Washington state to try his luck in commercial real estate but never could get his hands in quite all the way, and he didn't understand why his wife had left, still didn't know how to do things by himself.

Wayne laughed at Warren. "I'd have left you too," he said.

Wayne liked to tease, and Wayne was blunt. Any man who came into his house with a ghost, Wayne would find out about that ghost and make fun, try and get the guy past it.

He thought of his own divorced days in Washington. Sharing an apartment with a friend and cooking on a camp stove because they didn't know how to operate the real stove, putting Spic n' Span in the washing machine. Then a couple of years later he'd remarried, a Western woman. Barbara had grown up on a ranch and her grandparents had come out in a covered wagon. She and Wayne had moved to Alaska.

As a young man Wayne had been told that his loud laugh would keep him from promotion to the head office of Great American Insurance in Cincinnati, and he'd accepted that. He was too rawboned for Cincinnati, he was an outdoorsman and a salesman—dogs, children, grandmothers, he could get them all, Alice said. Now people said he was the most successful commercial insurance underwriter in the state. He was vice president in charge of Alaska, he worked in a building on Third Avenue.

Anchorage was booming. Wayne thrived in the flows of wild and woolly pipeline workers through the city. Nine weeks on the tundra, two weeks off. Brothels and bars were open all night long, and when someone complained about drunk drivers on the streets, Wayne laughed at him. "What do you think people are doing out then? Good people should be home at ten o'clock, and the police should pull over anyone who isn't drunk and arrest them."

Then the symphony came and Wayne worried that the civilizers had arrived.

There wasn't enough housing for the pipeline workers, so he and Barbara let rooms for men who needed them. Wayne was generous but Wayne could be hard. When two tenants asked for help muscling a refrigerator into an apartment, Wayne told them to wait, he was watching television.

"I'm watching the game. Everyone can sit and wait."

People cursed Wayne and said they wanted nothing to do with him, he was so hardheaded, but he laughed at himself, too. He knew his own ghosts, he said, he didn't have secrets from himself.

"You drink, Gardner, we smoke pot, it's the same," said a fellow who wanted to bring pot into the house.

"I may be wrong, I'm not going to change."

"Why don't you try it?"

"I drink heavy and I smoke three packs a day, do I really need another vice?"

You couldn't stay ahead of Wayne, you couldn't win.

The next morning Wayne was up at 6:30, an hour and a half before sunrise, cooking bacon and dicing potatoes.

"Wake up, Warren. It's cold out and you're getting grease and cholesterol for breakfast."

A pound of bacon and six eggs cooked together, no pouring off the grease, and Wayne gave Warren a primer on moose. Just walk up river a little ways, then cut in a half mile, you're guaranteed to see one. The bulls were in rut. They have few natural predators, and when they're in rut they're crazy. The bull will defend his harem. He'll run himself ragged chasing off the young spikes.

Wayne headed south. He liked the feel of the frozen tundra under his boots. When the tundra wasn't frozen, it was like walking on a mattress. An eagle sailed over the river.

Then he heard a shot, and a thud right after.

It was an hour before Wayne got his. He was a crafty hunter. He found a spot that moose liked, a sunny spot by a frozen marsh, and lay down to sleep. After that big breakfast, he felt sleepy. And anyway, going to sleep was one of his tricks. A hunter walking around scared animals off. He liked to sit down and let the animals stir around him.

Then his eyes opened, and he looked around to see what had woken him.

Several moose were across the marsh, 150 yards off, maybe closer, a bull and three or four cows. Wayne got to his feet, and the bull snorted, then started toward him.

He balanced his shoulder against a tree and put his eye to the scope. He could hear the moose breaking through brush, he waited for it to come into range, then it was fifty yards off and he squeezed on the Weatherby, and the moose crashed forward, and that was all.

The cows were gone. He started across the marsh. Dry yellow reeds poked up through the ice. The marsh was frozen deep enough to support a man but it had given out where the moose had gone down. Its right eye was gone, it was planted with its chin in mud. Fifteen feet away the smell hit Wayne, and he was nauseated. When a moose was in rut, it rolled in its own waste.

Wayne walked around it. The rack was over five feet wide, the moose had to weigh over 1,200 pounds, the biggest moose he'd ever killed.

And a devil to turn over. The left front leg was planted deep in icy mud. Wayne cut an aspen stake and drove it into the ice a few yards away, got a rope on the right hind leg and tied it off on the stake so the hind end corkscrewed over in the sun, exposing the moose's belly.

The smell was too much. The fur was matted with mud, shit, moose sweat, aroma to a cow maybe, but Wayne had always had a weak stomach. He got out his ten-inch knife and took a deep breath—turned, vomited—

Then he had the belly open and shoved his bare hands down through hot guts. It was so cold, he warmed his hands in the viscera then dragged them out on the swamp and vomited again, the rest of his breakfast, and cut the guts away from the esophagus, braced the abdominal cavity open with a stake so the meat would cool, before heading back to camp with the liver.

It was nearly one o'clock. It would be dark in two hours, he'd need Warren's help to get the moose out.

An eagle wheeled in the sky, the same one, he imagined.

Warren was in a mess of trees. Just 500 yards from camp, his moose had tumbled down into a slope of pines, a small moose, maybe 900 pounds, wedged with its head at a strange angle. Warren had gutted it right there. They quartered him and dragged the quarters back to the sandbar, using the moose's legs like handles. Wayne went for his lifter, and they had all the quarters up on the meatpole ready to be boned out and Warren was grinning in spite of himself.

"First moose."

"It's not huge, but it's a good-looking moose," Wayne said, then heard the plane.

A white Cessna was coming in low from the south.

"Doesn't that look like a plane from Harold's?" Wayne said.

It came low scoping the sandbar and went past them and made a wide turn to come in.

"What the hell," Wayne said. "I told him not to come till day after tomorrow, Warren, you heard me say that—"

"Said you'd set out the space blanket, red side up."

But the space blanket was packed in the tent, and the plane came in low over the river and kicked up sand on the bar. They got to it, and the pilot already had the door open.

"You're not supposed to be here till day after tomorrow."

"You're Wayne Gardner, right?"

Wayne didn't recognize the man.

"That's right."

"I'm sorry, they sent me in with some bad news. Your daughter's died."

"My daughter? That can't be. My daughter is in Tonga."

The man shook his head. "I'm sorry, I don't know anything about that. All they told me is go in and get Wayne Gardner, his daughter died."

Everything had stopped, everything was still.

"My daughter—"

The pilot stared at him intently. "That's what they told me."

Then it seemed to Wayne that the man was utterly serious, and that Wayne could slip away from his life right then like a brown leaf from a tree, his connection was so flimsy.

"What did they tell you? How'd she die?"

"I don't know any of that. I'm real sorry to have to do this. I'm supposed to come get you."

"Wayne, I'm sure sorry," Warren said.

Wayne turned away on the sandbar. He walked away from the others toward the river. There were ice floes in the river and it seemed they weren't moving. He tried to tell if the ice floes were moving or not, if time was going on or staying right where it was, then his body was like a block of ice below him.

Debbie is my daughter I don't have a daughter—

Something heaved upward in him and cracked under his throat. No man should have to do this, no man in the world. Then he heard his breath and he turned back and didn't show anything, but he could feel the men were afraid.

"Wayne, go on now, I'll take care of everything," Warren said.

Wayne didn't look at him.

"I'm not leaving you out in the middle of the wilds in this kind of cold, Alaska's no place to be alone if you can help it."

"Just go, man. Go and find out about your daughter."

Suddenly he did not care for Warren, couldn't waste energy on him.

"Warren, you listen to me. If you broke your leg—it's colder than hell out here. There's not many people that have spent any time in Alaska by themselves."

"Grab my moose and we'll leave the other."

Wayne's head shook from side to side.

"I'm not doing that, leave an animal to rot."

"There's laws against that," the pilot said. "You could lose your license."

"I wouldn't do that out of respect for the animal. Kill a moose, you owe something to that moose."

Then what was it about a person's eyes, he couldn't hide from the men's eyes. The men's stares were blinding.

"You come get us tomorrow."

"Leave me here, Wayne, I'll be fine," Warren protested. Now Wayne simply ignored him.

"Come around two, we'll be ready."

So the pilot was gone, and Warren said he was sorry again, he had a daughter himself.

"What was her name?"

He mumbled something back but he wasn't about to start talking about Debs, not to Warren Fortier.

Somehow it became dusk.

Warren roasted the moose liver on the Coleman stove, he gave Wayne scotch and Wayne dropped cigarette stubs in a can. His daughter was in the Peace Corps, he said. She had studied at Washington State University and gone into the Peace Corps. After that there was nothing to say. Men cried with women, they didn't cry in front of other men.

Night.

He walked away from camp, his boots scraped ice and gravel, still he did not feel a connection to the world. Didn't feel the tundra, only saw the lights in the sky. Those lights. They came down like a ghost, then they were gone. They came down and Wayne felt like he could touch them right on top of himself, and gone again just like that.

Your daughter's died. The words came down again and again and floated away. He wondered if he could keep those words away, hold them away from himself, but they were like the auroras, they came down whenever they chose.

He'd never governed her, she was a Gardner like him. No one could govern a Gardner.

Debbie going off down the street by herself and him running after her and spanking her right there on the street, what was she, two?

Debbie putting on six pairs of bloomers to visit the neighbors before he and Alice got up, a year and a half old, in bloomers with ruffles on them, getting the chain free on the door and going up the apartment stairs.

Debs at eight months crawling across the floor to his gun, and him saying No! The lesson he'd prepared . . .

Backwards, backwards. How far back would he go if he could to do it over, all the way back. Give me that chance, as far as you want.

The delivery room, studying his watch to Alice's contractions, Swedish Covenant Hospital and three other couples in the room and all the men had been in the service like him and the guy next to him with his feet up on the bed reading *I the Jury.*

Our eighth, the guy said, grinning.

My first.

Why not him, with eight?

Then Alice was in pain, the baby's coming, and Wayne went out in the hall and screamed at the nurse, Get in here. Mr. Gardner please, don't worry about it, I was just there, it's not due yet. But Alice was right, she was always right, there was a rush and Debbie had come.

Debbie on her own time, always. Debbie going to Tonga on her own. Debbie doing what she wanted. Debbie turning her fury on Wayne.

What had they been arguing about? Where had he been for Debbie? The white lights came down like a ghost and went away and came back. Nowhere is where, nowhere nowhere nowhere.

14

The Sendoff

The Anglican church was the city's prettiest, a delicate stone building angled in its courtyard to face the rising sun. Dwarf palms crouched outside the front door, bougainvillea spilled over the gate in jarring pink and orange, and at dusk that Friday, bicycles were heaped everywhere along the walls.

"We spend our years as a tale that is told. No one knows how their story will go, and once told, no word can be changed." Bishop Halapua had not known Deb but he said a few words about her. She had come out to help Tonga, it was such a pity that she had died so young and so far from her family. Then for a moment the bishop addressed the volunteers in the congregation, and judgment came into his voice. "We appreciate all the good you bring us but we have to be aware of the bad, too . . ."

And the volunteers resented that, felt it as a blow to their already bent shoulders.

Frank went to the front of the church and with the page trembling in his hands read what he had written. He started as drily and factually as he could.

"Deborah Ann Gardner. Age 23, home, Steilacoom, Washington.

"Debbie came to Tonga in December 1975. She taught science and home economics at Tonga High School. She was hoping to transfer to Ha'apai for the next school year."

Then Frank looked down, doing all he could to maintain his composure.

"A beautiful, vivacious, outgoing person who was full of life—her eyes, her laugh, her smile were her tools of communication. She used them freely, to anyone who came to her—"

A Tongan woman with a bell-like voice cried, Hymn 19, and voices filled the church and it was as if everyone heard these words for the first time—Amazing grace! how sweet the sound—and the people who had been closest to Deb rushed past the others to get outdoors, to the cover of darkness.

The rift that had formed between the Peace Corps and the Tongans widened that night. Some volunteers went to a place called Joe's Hotel to eat, and one of the volunteers had on a lei, and a waiter came up and rudely took it off his neck and said, "You are in mourning, you should not wear this." A drunk walked up and down Sālote Road outside, shouting, "Tongan men do not kill Tongan women, Tongan men do not kill women! *Pālangi* men kill women!"

Frank and Mike Braisted stepped through the wide front door of the Tonga Club and the place went dead silent. They walked up to the bar and they could hear a pin drop. A man twisted around on his stool to study Frank.

"Where's your fucking knife?"

"What is this, Siaosi, you know me, I'm Frank, I've been here three years—"

But others stared balefully at the pair, the *swipi* players put their cards down, and Siaosi said, "How could you do this to us? Please leave."

Many Tongans were angry. They were a period on the map, the smallest monarchy in the world. They liked their reputation as the Friendly Islands and had given the Americans to understand that they were guests who should be well-behaved. Now look what had happened, the Americans had brought out a choice piece of culture to Tonga and one that was likely to cast a shadow over the Kingdom.

There had not been a murder in Tonga in seven years, officially anyway, and the Tongans did not think of their own bloody history, when Nuku'alofa was the seat of a South Pacific empire and smiling warriors dashed out the brains of their brothers with ironwood clubs and then chased the wives down the sunny beach, the women wailing for mercy for the babies in their arms till every skull was broken, the little ones too, and the sea lapped blood from the foreshore.

The morning after the memorial service it was market day, and two volunteers bicycled into town together in the King's Road and a Tongan man called after them, "Tenisi, Tenisi," and the call was taken up. "Tenisi!"

You are all Dennis.

A new volunteer looked around for mangos in the market, and a farmer idling against the wall hissed, "*Tāmate'i*." Killer.

It was Peace Corps preference that a staff member bring a body back, someone responsible who could answer the countless questions that were waiting. "I brought them out, I'll bring them back," one country director put it. But Tonga was a small program in crisis, its existence seemingly threatened by the case, and Peace Corps was not about to part with one of its two full-time *pālangi* staff.

So a volunteer would go to the Gardners, and Mary and Mike Basile had decided on Emile. Even though Vickie said that she wanted to go, she was Deb's good friend, and others questioned the choice. Why Emile? Emile was close to Dennis, we never liked Dennis. Why not someone grave like Jackson or poetical like Frank? Emile is a cutup, will he tell jokes? Still Emile was chosen, and was honored to fill the role. He scraped together a funeral wardrobe, his scratchy striped wool pants and a pink tie, a blue blazer borrowed from Mike Basile that was tight over his shoulders, and prepared to leave Saturday.

That morning Deb's hut was still a horrible mess, and Emile went to the police to get permission to clean it out. Lolo Masi helped him, a small bearded volunteer, and Greg Brombach, too.

They carried bloodied mats out into the yard and burned them, poured buckets of water on the cement floor and splashed the walls, swept the bloody water out with a broom. They kept finding spots of Deb's blood they had not seen before. Sprays had reached the farthest ends of the room, and a bloody handprint streaked down the southern wall, the fingers leaving long tracks.

A boy stood at the low barbed-wire fence, scared to come any closer. "'Emili, 'Emili, 'Emili, 'Emili," he called in the drumming Tongan manner.

Then Emile turned. "Lapalani."

"'Emili—*kovi*."

"'*Io, kovi aupito*." Very bad.

They had given away her food, now they sorted clothes and her handwork, trying to figure out what should go home. They made the place look nice, put her last shopping basket made from a woven green coconut leaf on her low table alongside her parasitology text and biology books. They stacked her books. Deb could read a book a day, and Brombach made a pile of them and put all her letters in a box to go back home, and in doing so found a sheaf of poems. No one had even known that Deb wrote poems. The men were torn about whether to look at them, but they did. There was one she had written lately, about a man. Emile took it to give to Frank.

Emile was making the bed when he came upon a balled-up photograph that froze him. It was the picture of him and Deb at the airport. Deb was in her baggy clothes, and Rich Dann had his nose against her neck, and she was smiling for the photographer.

Then it had gone up on the Peace Corps wall, and now here it was, crumpled, and Emile thought about what might have set Dennis off.

"Emile, what are you going to say to her parents?" Brombach said.

Emile humped his shoulders. "I don't know. I just don't know—"

Then Brombach found her gold cross on the floor, chain broken. Who had cut the chain? He put that in the box of things to go home and piled her travel brochures for New Zealand, too, as if he might still respect her plans. Then her passport slipped to the floor and two photographs fell out. Deb had a big smile and a fresh gleaming eager haircut, and Brombach thought of meeting her in San Francisco and had to wander out into the rugby field and cry.

Deb Gardner had made food for him, Deb had stuck her finger hard in his ribs making fun of his broken Frisbee. How did you break a Frisbee? Brombach had always had a crush on her, never voiced it.

When they were done, they went over to the police station to see if they could get Deb's backpack to go home, and they both had a look into Dennis's cell. Brombach had wanted to beat the shit out of Dennis, now he felt pity. Dennis didn't look like Dennis, but shrunken and mad. Emile asked Dennis if he wanted him to call his parents when he got home. Dennis said no, he should call senators and get him out of this place.

Out the back of Vaiola was the morgue. A bunch of girls from Tonga High had sat up all Friday night on mats on the floor, wailing and singing songs alongside the refrigerated unit with Deb's body inside it, as Tongans always did when people died. Because death was a lonely journey and the living should provide as much company as they could.

Losimani, the nurse married to the Peace Corps administrator, ironed one of Deb's dresses and pulled it down over her cold body, brushed out her hair. Then Losimani fretted that Deb was going home with bare feet, and what would her mother think if she came home barefoot? That she had not been appreciated by the people she'd gone out to help. So someone went to the market for new pink socks, and meanwhile Mary knotted one of her own scarves around Deb's neck, concerned about Deb's mother seeing those cuts to the carotid and jugular.

As at all remote Peace Corps stations, an object of ghastly providence was lodged above the ceiling joists of the Peace Corps office beside the lagoon, an aluminum casket suitable for tropical conditions with plastic bubbles on the inside to hold a body out from the metal, and the big driver Pila now got it down and Deb was put inside it and the casket was then placed in a wooden crate for shipping purposes. Four copies of the death certificate were obtained from the health department. Mary's name was on the certificate. She had identified Deb's body after all. She had overcome her objections to the police investigation, that much anyway, and gone to the hospital in the presence of an officer.

Puloka gave the cause of death as massive hemmorhaging from stab wounds. He listed three fatal wounds. The two to the neck had been preceded in his opinion by the deep wound in the left side, which had required great strength. The knife had broken right through the rib and nicked the kidney and cut the descending aorta. The laceration to the aorta was about three-eighths of an inch, nearly a third of its width. In such a major vessel inches away from the heart, that was huge. Deb may have lost half her blood in fifteen minutes from that cut alone.

The crate went into the Peace Corps pickup truck, with Tongan and American flags on top of it, also bird of paradise flowers, and at three o'clock outside the hospital, a long parade of cars and buses formed behind the pickup. A volunteer told Mike Basile not to let Mary drive. She was driving erratically, he reported, she had almost hit a child, so Mike took the keys from Mary, and the wheel. Mary wore dark glasses, a white dress, a wide-brimmed hat. She had not been sleeping, she was drawn.

Tonga High students in their maroon uniforms lined the King's Road and bowed their heads, an honor granted to royalty, and at Fua'amotu airport the girls formed rows and sang hymns. Their faces were swollen and puffy. These were the girls Deb had taught to sew stuffed teddy bears and to cook macaroni and cheese, girls for whom she had sketched out the circulatory system on the blackboard in Form 6 biology.

The Air Pacific plane came in from New Zealand, and the students made two opposing lines out to it to see Miss Gardner away.

Pallbearers brought out the casket in its crate, a crowd of Tongan and *pālangi* men. Mike Braisted stood off to the side thinking ironic thoughts—pomp and circumstance, not quite the way that Deb had planned on leaving—till Mark Stiffler said, "Hey, Mike you get a hold here, too."

"There's nowhere to go."

He went on the front, and when they got to the plane the flags were removed and the trouble began.

Air Pacific employed a short-haul plane, a BAC-111, whose cargo door could not accommodate the wide wooden crate. So a couple of men got inside and a couple outside and they tried every way they could to jam it in. They tilted it, and shook and chattered it as they pushed, then two or three of the bigger Tongan cargo handlers stood outside and torqued it, trying to rack the wood, before letting the thing drop to the tarmac.

"Why don't you dropkick it while you're at it?" Mike Braisted snapped, walking away.

But Emile said, "Look at it this way, Deb would be laughing if she were here."

Then it was decided to strip the crate, and the men began to pry the framing off with a hammer and bar right there when the pilot came out and said he couldn't fly the casket if it wasn't in its crate, that was regulation. He became insistent, and they stood out on the tarmac arguing as Deb's body lay there.

The plane had to go, it was due in Suva before nightfall.

At last it was suggested that the casket be carried in the passenger compartment, and as there was room, the pilot agreed. Five seats were removed toward the back of the plane, and the crate brought through a passenger door, and the door closed.

The plane taxied, Rolls Royce engines whined, it hurled itself down the runway, glinted like a needle in the western sky.

Emile sat with his left elbow supported by the casket, and the stewardess called him Honey and brought him a glass of scotch. He realized he might have to say something about Deb and got out a piece of paper to write some things down.

"What will I miss the most about Deb? The sound of her voice. Her laughter. Her smile. Her honesty. But most of all I'll miss the close relationship that slowly evolved between the two of us."

He read that over and scratched out the word "relationship" and put in "friendship."

He thought about Dennis's face earlier that week, in the days after Emile had taken Deb home from the Dateline dance.

Three times Dennis had ridden by him on his bicycle and spat harsh words at him. Monday morning Emile was bicycling to school and Dennis came the other way and bore down on him, said something very hot at him, or very cold. "What?" Emile said, but Dennis rode on by. Then that afternoon, Emile was sitting outside Tonga High when Dennis came curving past with steam coming out of his ears. "You fucked her," he hissed, and kept on going. Emile wondered if Dennis had been hiding outside Deb's house Saturday night, spying on them.

The third time Emile was eating at John's Takeaway, the restaurant in the King's Road downtown, and Dennis rode up and stared at him. Some of the intensity was gone but he was still angry.

"You slept with her, didn't you?"

That was private, that was none of Dennis's business. "Believe whatever you want," he said. "Why don't we have a drink at the Club?" But Dennis rode away,

Emile considered the page and crossed out "friendship," put back "relationship."

·   ·   ·

Frank couldn't bring himself to go to the airport and see Deb leave. He lay down at the mosquito lab and slept for the first time in days.

When he woke up that night, it was dark and Deb was gone, and he went back to the Tonga Club and this time got in, and a lot of other volunteers were there getting drunk. None of them knew quite how to handle this. It's the heaviest thing that ever happened to me, Brombach said, it feels like my cousin killed my sister. While Jon Lindborg noted in his journal that life had narrowed down to three things— "Loneliness. Bitterness. Tears."

Jackson was comforting. Jackson was going to be a nurse. "Listen. All the pain you're feeling about Deb, just remember—she's not feeling any of that anymore. Wherever she is, she is past that."

They talked about Dennis. A lot of them said they should have seen it coming, that Dennis had gotten weirder and weirder and they were either naive or they'd put off doing anything about it. Lindborg said he knew Dennis was strange when he saw him carrying that extra tire, Brombach said it was when he brandished his big knife at a dog.

"I always felt he was sick," Mike Braisted said. "If something hadn't happened now it would have happened later, maybe not to the same degree—"

He just hoped that Dennis had some friends to visit him in jail. Deb's friends weren't about to do that. Many of them didn't care what happened to Dennis now, he'd lost his humanity. They didn't want to see him or hear about him again.

Among some of them, the view took hold that Dennis was an evil genius. The methodical planner who planned everything he did and generally succeeded had planned to try to get Debbie, but she didn't want him, so they speculated that he planned her destruction backward and forward, even her wounds. "He was as close to genius—you know as well as I do how close genius is to madness," Brombach said, "and he was always complaining about there not being anyone he could talk to."

There was one story about Dennis going to the beach with a friend and wading out into the water, talking about how the moment of truth was when you had to get your dick wet. Dennis said he was able to control his dick, he could use mind control to keep it dry. "Look at this—" And the friend looked over and

Dennis had caused himself to get an erection under the bathing suit, to keep his dick out of the water.

They left the Tonga Club at midnight, and Lindborg and Stiffler had only one bike between them, so they rode double down to the lagoon, laughing and falling as they went. Stiffler was still laughing when he climbed the steps of the mosquito lab but as he went through the door he thought of bicycling here alongside Deb one day when Deb was wearing pigtails and they both had coconuts on their bikes, and he started sobbing.

Frank held him. "Hey, Mark, is this the first time you've cried? Man, you've got to let it out—"

Then Mark went home, and Mike Braisted heard him crying through the thin wall. Mike felt guilty. He wondered if he'd been too distant to Deb lately, too self-involved. He had begun courting Jackie, another volunteer—indeed, the volunteer he would one day marry—and when he read over his last journal entry about Deb, it was painful under the circumstances. "I'll still try to see Deb, but old fickle Miguel seems to be on to a new person."

He had tried to see Deb that week. Last Saturday night he'd tried, but she left the dance with Emile. Monday he'd tried, and she wasn't home—out with Dennis. Tuesday she was busy again, with Stiffler and Lindborg. Wednesday he'd been busy. Then came Thursday.

Emile had given Deb's last poem to Frank, but Frank wasn't sure the poem was about him, so he had passed it on without comment to Mike. Now it hurt Mike to read it.

> *He may not write you every day or even call*
> *And yet that does not signify he wants to be alone*
> *He may be very occupied with other things in life*
> *And not concerned at present with the prospect of a wife . . .*
> *So why not just accept him for the friend he tries to be*
> *And make the most of every time you share his company*
> > *You know not when the rains will cease.*

It had rained just about every time they did something together, Mike thought.

People had come to Mike to express their sympathy as if he were family. He'd cried in front of them. He wondered when Frank was going to let loose.

Most times in life, Frank's heart was on his sleeve. Everyone knew what he was feeling. This was different. He couldn't really begin to talk to anyone, so he wrote a poem about Deb, though he found it so painful to write her name that he didn't, but stuck with a nickname he'd had for her, "Snoopy." Then it was so painful to write the words "she" and "her" that he used male pronouns.

Hey did you hear?
Snoopy left yesterday.
Just picked up and left,
didn't say good-by to anyone.
Saw him for the last time at the zoo,
playing with a houseful of monkeys. . . .
Someone told me he was last seen
marching up the road out of town
with the charisma of a pied piper.
All his friends and a few curious onlookers
trying in vain to change his mind,
Maybe trying in vain to change it around in their minds.
Goddamn
Sure gonna miss him.
Nice having him around
Even for so short a spell,
But that's just the way he was—
Stops some place
Mixes everything up
lets it simmer for a while
And it comes out tasting
Just about the best

You could imagine . . .
I kind of know deep down
That we will all be getting together
later on.
Only we'll have to find him.
Because Snoopy isn't going to do any backtracking,
In this direction, at least.
Wasn't his style. . . .

15

A Vision in the Cathedral

D ennis—Dennis—how are you?"

Within a couple of hours of getting the news Friday morning, the Bird Lady went to the hospital. The police would not let her into the room, still she called out to Dennis through the door, and he responded, groggy from the overdose of Darvon.

"Barbara!"

"Do you need help?"

"I need a psychiatrist."

The policeman at the door did not even know that word, psychiatrist, but he put a stop to the conversation at once. He and another constable took the Bird Lady aside and asked her several questions about Dennis. It felt to her like an interrogation, as if they regarded her as an accomplice. They let her leave a note for Dennis, then directed her to go.

That night she needed Valium to get to sleep. A hundred volunteers seemed ready to stone Dennis in the King's Road, and Kalapoli Paongo, the principal of Tupou High, had opened the assembly that morning with the statement that Dennis was guilty of murder. Dennis was her friend, he was one of her favorite people on the island. Now he was isolated and everyone was condemning him.

She woke up the next morning to the usual chorus of roosters and honeyeaters, but nothing else was normal, and she started to cry. In the previously normal world Dennis would never have killed anybody. In this new world—

She stood at the window trying to distract herself by watching a neighbor try to get a mango down from a tree by throwing rocks and sticks at it. Fifteen minutes the girl spent, hit it once, but it didn't come down. She shifted her attention to a lower mango, and then back to the higher one.

The Bird Lady told herself it wasn't true, Dennis could not have done it. Yes, Dennis was strong and would defend himself if you came into his house with malicious intent, but he was like one of her brothers. For all of his rough exterior he could be remarkably gentle. Even when he threw the students out of his classroom for failing to bring in their potted phototropism experiments, the Bird Lady had understood.

Anyway, from the sound of it, it wasn't Dennis's type of murder. He did things well, this was a mess. She was sorry Debbie was dead, but she had never known Debbie. Her first thought had been, "Poor Dennis."

On Saturday morning the police brought Dennis back to the jail, and two volunteers near the market saw him through the Land Rover window and he raised his bandaged wrists to his face and banged his forehead with his hands as if to say, Whatever got into me? Dennis had never been embarrassed. Now Dennis was embarrassed.

He was in some crisis of knowing. One thing that could be said about Dennis before October 14 was that he had intellectual confidence, he knew himself to be smarter than other people. Since he was very young he had known that he was gifted, known it without his teachers having to say that to him. Now, at 24, a curtain had been pulled back on a part of his being that his brain had not contended with before. However logically and cleanly he had planned out his actions, he had failed to predict the chaos that had resulted, in Deb's *fale*, or his own. He had always possessed the hard scientist's hubris about psychiatry, talked about psychiatry as garbage science. Well, that was the first thing he'd done, ask for a psychiatrist.

A group of students came by from Tupou High, ostensibly to pray for him, but they forgot their prayers as they peered in at their former teacher, clad in police-issue khaki shorts and khaki shirt, his wrists bandaged, feet bare, glasses gone.

Dennis did not appreciate being a human zoo exhibit, Murderer, in a Third World jail no less, and he demanded his release. If he was not released he would starve to death. His first suicide attempt having failed, he now set about it more

Dennis-like, more methodically. He remembered that it took a man with no water five days to die. Majoring in biology was good for something after all. He would not eat or drink till he was released.

That day again the police turned the Bird Lady away. Dennis was a criminal defendant charged with murder, he was not to receive visitors. Paul Boucher and John McMath were more successful, Kalapoli helped the two men gain access. Whatever Kalapoli thought of Dennis, he was the principal of the most prestigious school in the country, and he had pull with the police on behalf of his staff.

Paul sat with Dennis for an hour, tried to convince his friend to eat. John brought a pineapple and a Bible, shook Dennis's hand. "Hullo, Dennis. I want you to know that you are in my prayers. I urge you to pray too, God can forgive anything." Paul spoke to Dennis about the sanctity of life. "Each individual soul has value," he said. "How could you do what you did?" Dennis was stunned, silent. Paul went on as if lecturing Dennis. The killing violated every important human principle.

Two others were also able to see the prisoner on Saturday. A lawyer was shown in, actually a former bookkeeper licensed to practice in Tongan courts, a dignified and sallow man called Tomi Finau, who listened quietly as Dennis stated that he did not remember much of Thursday night beyond a struggle at the girl's house. Then Dennis listened keenly as the lawyer described the possible sentences arising from different verdicts.

Premeditated murder: hanging or life imprisonment. Unpremeditated murder: life imprisonment or less, depending on circumstances. Manslaughter: fifteen years. Temporary insanity: open.

No, not a lot of daylight. As Finau went down the grim list, a third man sat in the cell taking discreet notes, a silver-haired man in a blue blazer.

The Crown Prince had called the chargé d'affaires in Suva, Fiji, first thing Friday morning, and the chargé had promptly cabled the Ambassador in Wellington. Now here he was. Flanegin had come out from Suva Saturday morning on the 8:30 flight and dropped into one office after another. He met with the Crown Prince and the government secretary and Mary George, too, and when he stopped in at the jail, the police got Dennis's Seahorse knife out of the cardboard box to show him. The chargé's business was done inside of six hours, and he went back to Suva that afternoon on the same flight that carried Deb's body away.

Flanegin was brisk, that was his signature. He could find his way around Paramaribo and Port Louis, he could find his way around Nuku'alofa. So he grabbed the 4:45 home like some commuter, and turned a number of elegant phrases in his report to the Ambassador.

> LITTLE TO ADD TO TELEPHONE CALL REGARDING KNOWN
> CIRCUMSTANCES OF GARDNER DEATH. SAVAGELY ATTACKED WITH
> WEAPON RESEMBLING FISHING KNIFE, BODY COVERED WITH
> WOUNDS. KILLING OCCURRED AT HER HOUSE AND LATER IN EVENING
> PRIVEN TURNED HIMSELF IN TO POLICE AFTER SLASHING WRISTS
> AND TAKING HEAVY DOSE OF DARVON, A TRANQUILIZER HE SAID TO
> HAVE BEEN REGULARLY USING FOR LAST 6 TO 8 WEEKS.
>   PEACE CORPS DIRECTOR MARY GEORGE DENIED ACCESS TO HIM
> BY SOMEWHAT CONFUSED AND OVERLY CAUTIOUS TONGAN POLICE
> WHO AGHAST AT FIRST MURDER INVOLVING FOREIGNERS IN LIVING
> MEMORY. PRISONER WELFARE ASPECT OF CASE COMPLICATED BY
> CONFUSION OF POLICE WHO HAVE NO PRECEDENTS TO GO BY,
> FEARFUL THAT THEY WILL BE ACCUSED OF ACCORDING SPECIAL
> TREATMENT, YET HIGHLY AWARE OF US CONNECTION OF PC.
>   PRIVEN APPEARED HEALTHY THOUGH MOROSE, DEPRESSED,
> UNFORTUNATELY ABRASIVE WITH POLICE.

The chargé had smoothed things out. Dennis was upset about his feet. The police had refused him socks, fearing that he would hang himself with them. How could he possibly do that? So the chargé pressed the police, arranged for a new pair. He had also worked on Dennis not to kill himself.

> FINALLY AGREED TOWARD END OF INTERVIEW TO DRINK WATER
> PENDING ARRIVAL OF PSYCHIATRIST.

Robert Flanegin was upbeat, he assured the Ambassador that matters were well in hand. The Tongans were allowing Dennis visitors, a psychiatrist would soon be there, and a psychologist, too, the former country director Dick Cahoon, now in Samoa.

Actually, matters were not well in hand. Nothing had been truly resolved by the chargé's lightning visit. The police would continue to be arbitrary about visitors, Dennis would continue to threaten suicide. But as quickly as he'd appeared, the chargé was gone, and Mary was again the highest American official in the Kingdom.

It did not matter that Mary had feuded with Dennis, that Dennis despised Mary and Mary had seemed to despise Dennis. All that was ancient history, now Mary's passion flowed toward him.

She had at first opposed plans for a memorial service. "Debbie is dead, there will be a funeral for her in Tacoma," she said, and Mike Braisted had needed to explain, "The memorial isn't for Deb, it's for us." Subsequently, she had urged volunteers to write to the Privens in Brooklyn and say that everything that could be done for Dennis was being done, and she was so lacking in sense as to include Mike Braisted in that appeal. He and another volunteer complained to Mike Basile, who assured them they didn't have to do that.

The crisis took on spiritual moment for Mary. The police refused to let her in to cell number 1, and then she learned that they had questioned Dennis without counsel being present. She became overwrought, and rushed to the cable machine to tell Washington.

WE BELIEVE HIS HUMAN AND LEGAL RIGHTS ARE NOT BEING HONORED.

PLEASE PLEASE SET ASIDE BUREAUCRACY AND CALL PANGO TO SEND PSYCHIATRIST IMMEDIATELY WE NEED HELP CASE UNPRECEDENTED.

Dennis was being prejudged. No one even knew the evidence. The community's judgment might actually kill him. Or if that didn't do it, the police abuse would.

Mary's Bible study group began deep prayer for Dennis, and one of that company, a New Zealander called Ruth Judge, initiated a fast to protest his treatment. Fasting was a means of pleasing God, glorifying God. The Apostles had fasted in order to intensify their prayers, for fasting demonstrated to God that a worshiper was humble, and was able to control her baser needs and demonstrate that she was a godly creation.

Mary's energies did not end the cynicism about her. People wondered what she had known about Dennis ahead of time and what complaints Deb had made as Dennis stalked her. A couple of days before Deb's death, Mary had had a run-in with Dennis. He'd gone into her office. The door closed and a Tongan in the outer office heard the sounds of an argument going on for several minutes. From the outside it had sounded like quintessential Mary, withering Mary, Mary who was firm and brooked no opposition. Maybe Dennis was still pushing to stay for a third year.

The voices inside the room rose and there came a thud that could have been a desk being pushed. A loud cry, a scream, and Dennis left, slamming the door. His face was red. He didn't look at anyone, but ran down the staircase to the lot, and a few seconds later Mary also stormed out.

On Sunday morning, a meeting of Dennis's supporters took place in the same upstairs room of Viliami Kapukapu's house at which the new volunteers were having language and cultural training during the week. A dozen people took cushions on the floor, and Mary held court, telling them about everything that was being done for Dennis.

The police had left him on just a bare cement floor at first, saying he could have nothing that he might use to kill himself, but bedding had now been supplied at Peace Corps insistence. The lawyer had been found, what expats called a bush lawyer. Tomi Finau was highly intelligent and the most experienced criminal defender on the Tongan scene. The psychiatrist from Hawaii was due to arrive in a day or two. It was important that people tell the psychiatrist about Dennis's mental state in the weeks leading up to last Thursday.

"Oh, but the police are being so hard and difficult," Mary said. "They have kept anyone from visiting Dennis."

Marie McMath interrupted. "John has been to see him, he went last night."

Mary's eyes glittered.

"He did? I did not give him permission."

"He doesn't need your permission, he went as Dennis's pastor." Then Marie's fury rose. "We don't need your permission to do things on the island—"

Soon after that Marie left. When she got home, she said to John, "What is a bike?"

"A bike? Just what it says."

"No, when a person is a bike. Someone at the meeting said that Debbie Gardner was a bike and others agreed."

"The village bike," John said. "An easy lay. Like anyone can get on a bike, you can ride it."

"I've never heard that before."

"Well, that is what it means and it's not a very nice expression."

Late in the afternoon of the sabbath, the Bird Lady went back to the jail and the police called a higher-up, a big light-skinned superintendent who came out and told the Bird Lady that much as he would like for her to be able to see Dennis, the police minister would not allow it. The Bird Lady surprised the man by saying that she didn't have to see Dennis, she only wanted to leave him a message, whereupon the superintendent softened and told her to stick around. Dennis's lawyer was coming by, the prisoner might be brought out of his cell.

The Bird Lady stuck around and stuck around, now and then looking at a bulletin board that said police were to check cell 1 every fifteen minutes because the prisoner was under suicide watch. The Bird Lady was not religious, but she went to church Sundays and recognized that her need to see Dennis had a biblical character. There was a family drama in her background. She had grown up inside a tight-knit family in a small town in Michigan. Her parents were upstanding members of the community. Then something had happened, she did not talk about it to others, it was too painful: a relative had done something wrong and been publicly accused. In an instant her family had been ostracized, and the legal process had been tortuous.

The experience had wounded her, and made her skeptical about authority and public opinion. So in addition to wildlife the girl began to study society, as an outsider.

The bush lawyer arrived, a thin man with aristocratic features, and was led back through the charge room to cell 1. The door opened, the Bird Lady saw Dennis and exchanged sheepish grins with him. Both Dennis's wrists were bandaged, and the rubber bands he normally wore on his wrist were gone.

As the cell door closed, the Bird Lady called out "Good night" with all her spirit.

The police officers were looking at her, and not as if she were some freak but with warmth, and it came to the Bird Lady that her constancy had made her into a romantic figure, the forgiving girlfriend. That had never been her role, Dennis and she had been brother and sister. In fact, the Bird Lady was involved with another volunteer, John, who'd gone back to his family farm in Iowa. Still, it did not hurt now to let the police think that she was Dennis's *mafu*. That way at least she might get in to see him.

That evening, Mike Basile called his boss to check in on her but did not find her at home, nor at the Peace Corps office, and so he got into his little Nissan Bluebird and drove around Nuku'alofa to look for her.

The streets were dead, Mary was nowhere to be found. Mike went to Sālote Wharf. Mary had made plans to go boating that day with her Bible friends, to get away from the crisis. But she was not at the wharf either. Mike searched the horizon without seeing any pleasure craft. It was after seven o'clock, boats were in by now.

Mike worried about his boss. He was moved by the fact that Mary was alone at the top, a single divorced woman with lightning temperament living too close to the sea.

He drove back through the city. The announcer came on the radio. "In Nuku'alofa Friday, one Peace Corps volunteer was charged with the murder of another volunteer." A3Z was keeping the names confidential for the time being, but it had broadcast the first news of the murder anywhere in the world.

Mike passed the sagging police station, and there was the Land Cruiser outside it.

He felt the tension as soon as he went through the door. Three or four large police officers stood behind the counter inside the charge room. Their faces had a bristling focus, but they did not see Mike coming in. They were watching Mary.

She straddled a bench, sitting in a way that Mike had not seen her sit before, not very ladylike, knees apart, her dress riding up. He could always tell when she was anxious because beads of sweat formed on her upper lip. Now there were beads of sweat on her lip and she was bent forward in an imploring manner, clamping Dennis's bandaged hands in prayer.

The prisoner looked down, bewildered, thin, and pale, and lacking his usual fence against the world, those dark glasses.

"Dennis! Dennis! Do not lose heart! The Lord is with you! He will not abandon you! You should not be afraid—Dennis—"

Mary had brought him a Bible, she had marked key passages for him to study. Her voice rose as she quoted.

"Remember, judge not that ye be not judged. Can you hear me, Dennis? Can you hear me?"

She inclined toward him almost desperately, and he nodded his head. "Yes, Mary."

And the suffering servant in the Old Testament. "He is despised and rejected by men, a man of sorrows and acquainted with grief." The police stood like soldiers, and Mary resented their presence, she spoke at times in a hushed manner to keep her meaning from them.

Mike went around the counter, waited till Mary saw him. "Mary—excuse me. I've been looking for you, I'm so glad to find you." Mary smiled gratefully back at him. "When you're done, can you please stop by my house?"

Twenty minutes later Mary's car came into the little driveway at the back of his house in Sopu.

Janice Basile got Mary lemonade. The three sat down in the living room, furnished in cheap Danish modern.

Mary began to talk, and grew so agitated that she rose and paced the room or

sat in another chair and rose again. For more than an hour she held their attention. The Basiles were startled into silence.

She had gone boating that afternoon with her Bible group. Ruth and Tanya
were very different. That was the thing about Tonga, expats must make do with
whoever else they found themselves with. Tanya Dobbs was beautiful, an English-
woman with dark hair cut fashionably at her neck, married to an architect who
had come out to help the government. Tanya was quiet, serious, cool, and worked
for the Red Cross.

Ruth Judge was short, roly-poly, loving, warm. More than warm, Ruth was
possessed of fervor. She came from a small town on the South Island of New
Zealand, and she and her husband Robin were sunburned people of the land who
had signed up in middle age to be Volunteers Serving Abroad, the New Zealand
version of the Peace Corps. The Judges were deeply Christian, but not in a con-
ventional sense. Disdaining churches, they studied the Bible and the life of Jesus
and felt that religion was no book, it was God calling a person to action in a flawed
world.

Ruth imbibed Jesus' teaching of loving-kindness. Jesus had not gone out to
help the strong, he had helped the sick, people who had difficulty managing their
lives. Ruth had compassion for the abject and no feeling at all for the proud, the
entitled.

Tanya and Mary were of a kind, cosmopolitan, but they too believed in the
power of prayer, and Ruth provided leadership.

Ruth's favorite stories were about clearing away egotistical judgment. Mary
Magdalen was an outcast, but Jesus did not care, he had mingled freely with the
prostitutes. Zacchaeus was despised because he did the Romans' work by collect-
ing taxes from the Jews, still Jesus had gone to dine in Zacchaeus's home. "Zacha-
eus is also a son of Abraham," Jesus said, and his kindness to the tax collector had
baffled the churchly, the people of status. Jesus had even met with soldiers without
rebuking them. He talked to one soldier about his personal life not his profession,
and cured the soldier's sick son.

Bible stories were living stories for Ruth. Dennis was a wretched man, the
Tongans were determined to hang him, he wanted to kill himself, and the Tongans

were keeping her friend Mary from seeing Dennis. This was insupportable, and Ruth began fasting and praying.

Now a deep reservoir of passion overflowed the dams of her being, she went sleepless considering what she should do, and called on the Lord for guidance . . . At times in life, even Ruth seemed to harbor a small doubt about her judgment. She was not a logical person, she was somewhat insecure about her analytical ability. Yet if anything, that insecurity led her to be even more intense when she had identified a cause.

On the sailboat, the women discussed the prejudging. The Tongans would do anything they could to convict Dennis, but what was known? What was proved? Nothing. Dennis might even be falsely accused.

And when the boat was fastened back at the wharf, the women went to church.

The Catholic cathedral stood a couple of hundred meters from the wharf at the eastern edge of Nuku'alofa, a clumsy building in dark gray coral block, Malia 'Imakulata, the Virgin Mary. In airy Tonga, the architecture was almost medieval. Two square clock towers rose on either side of the nave, and a white statue of Mary was perched high over the front doors. Inside it was rather dark. A worshiper might almost feel entombed. Tapa cloth draped the pulpit, a crude wooden crucifix rose over the back of the altar. A series of livid paintings in pink plaster frames hung at angles from the walls to depict the Stations of the Cross. Christ stumbled, a helmeted soldier pushed his spear into Christ's back.

Dusk had come, the church was empty and cool.

"I was looking at the statue of the Blessed Mary, and I prayed to Jesus that I could see what had truly happened," Mary told the Basiles. "I prayed as hard as I ever have. And you know what happened—Mike, you'll think this is very strange, but it happened.

"You know the curtain that hangs behind that statue, the dark one? Well, that curtain moved. And there was no wind or draft at all in the whole church.

"Good God, it's hot, isn't it?"

She seemed to mean the Basiles' house, Jan got her a hand fan.

Mary's eyes had gone to the curtain and a face appeared clearly on the wall above it, the bland face of a Tongan, and not just any Tongan, but the superintendent she had dealt with during the spate of break-ins, a big light-skinned man, an oppressive officer who had been at the hospital when she had identified the body.

Mike broke in. "Maeakafa."

"It was Maeakafa's face, Mike, it wasn't Dennis at all. It was Maeakafa, I swear."

Then Mary's course had been clear, to go to the jail and comfort Dennis and convey her understanding that he had been falsely charged. When she got there, who should be inside the police station but Maeakafa.

Siaosi Maeakafa Aleamotu'a was number four in the hierarchy, a superintendent under Faka'ilo, the big phlegmatic man with sleepy eyes who had encouraged the Bird Lady to stick around earlier that day. Maeakafa had royal blood, he lived in a big house in the old city.

Mary had not come out and accused Maeakafa, but she had disputed with him over who was responsible for Deb's death.

"Priven definitely killed her," Maeakafa said.

"How can you be so sure of that?"

"Because. I have been there to see for myself."

"When?"

"That night."

"You have been where?" Mary turned the tables. "Where have you been?"

"I saw for myself at the girl's *fale*."

"You were at Debbie's house?"

Maeakafa nodded. "Yes. I saw."

Now moving around the house, Mary said that she had sensed Maeakafa's guilt in those words. "Don't you see, he was saying that he had seen the actual act committed, not how had he seen that?"

Mike's face felt hot. He was religious himself, he was Catholic and had worshiped in Malia 'Imakulata. Now he had a frightening sensation of being on his own far from home, on the most peculiar of stages, and the monster of the murder as he imagined it was now playing out its role completely, lurching about heedlessly, gorging itself on the lesser players.

"But Mary—remember our conversation with Paul Boucher."

At four or five in the morning, they had gone with Paul to Dennis's house in Longolongo and Mike had asked Paul whether he thought that Dennis had killed Deb.

"Paul said he had no doubt that Dennis had done it," Mike said. "And Paul is Dennis's best friend and colleague."

"Paul can't know for sure," Mary said. "Anyway Paul is too negative."

Mike stopped trying to change Mary's mind, though he said that he didn't want her to drive home by herself, he would drive her. The director's house was a half mile west of his.

"Mike, can you look through the house first?"

"Mary, he's locked up," Mike said hollowly.

"Couldn't you just walk through the place to make sure? It won't take long."

So Mike went through every corner of the house, the closets, too.

"It's all clear, Mary. Do you feel all right here?"

"Check the attic too, Mike—please? Sometimes I hear things in the attic."

A ladder went up to the attic, to dusty old boxes and a couple of pieces of furniture. Mike looked there as well, and then the next day he called Jack Andrews in the Solomon Islands and said, "Mary is hallucinating," and Jack said that he was on his way back.

16

Funerals

They had not talked in six years, but when Wayne came down the ramp of Seatac Airport, he said, "Hello, honey," and Alice said, "Hello, Wayne," and Wayne hugged Alice and held her. Then it didn't matter how long they had been apart, it was just Alice and Wayne together.

"I am so sorry," Wayne said.

Alice's eyes filled but she didn't cry, and Wayne felt for her. But Alice pitied Wayne more. She had never seen him looking so torn up.

She gave Wayne the news in a cool way.

"Debbie's body is coming in late tonight. A man called Emile is coming with her, he lived next door."

They sat at the bar there on the second floor, Barbara at Wayne's side and Justine next to Alice, having stiff drinks and talking about anything but. Justine did more of the talking, Alice was quiet. She was in a haze, she let others' words go by in a blur and let her eyes settle on small details. Wayne's glass. Watery scotch. When they'd first met he'd drunk scotch and soda. Men drank scotch and soda then. Bogart drank scotch and soda, and Wayne drank scotch and soda. Who was that movie star Wayne was in love with, no one has heard of her anymore, Debra Padgett. Then when Debbie was born they named her after Debra Padgett and he was in the Air Force and he switched to scotch and water.

Wayne's way was to talk. "Did you hear how they got me?" He told about the plane coming in after he got his biggest moose ever, he talked about packing the animal out over the next day in fifty-pound packs, nearly a mile across the tundra, then its antlers too, five feet across, more than fifty pounds.

No one would know it wasn't four friends having drinks at the airport till Wayne got a second and his voice changed and he said, "Alice, is it true what Barbara heard, Debbie had been stabbed many times?"

"Yes," Alice said, that is what she had heard too.

But they'd caught the boy who did it. "Don't worry about him, he's in jail," the Peace Corps said. The police had him there in—and Alice couldn't say Tonga, it was too painful even to hear the name of that place—the South Pacific.

A lady from the Peace Corps was coming out to Seattle the next day. Roy and Agnes were coming too, Wayne's parents in Richland. But the Johnsons were not coming from Wisconsin. Alice's mother Marie was 75 years old and too upset to come.

Wayne and Barbara got a car and drove to the Doubletree, then Barbara went to bed, but Wayne went down to the bar and had another scotch, or three.

He couldn't sleep. He'd begun a task that was going to take him a long time, maybe so long he'd never come to the end of it, thinking about what a bad father he'd been.

At the airport Alice said how much Deb liked teaching science in Tonga. Teaching science, Wayne Gardner had not even known that. He didn't know what was going on in his own daughter's life.

Because he had trusted her so much.

That was just an excuse. Still, it was true. He'd trusted Debs to take care of herself. It was Craig he'd worried about. He'd thrown Debbie's high school grades at Craig, because Craig was not as mature. Then she said, You give more attention to Craig.

Well, that's why, Debs—because you were more capable.

That wasn't an answer, it was a mistake. Mistake number, number what? Number 10,000.

In high school she came to him and said he loved Craig more than he loved her because he went to Craig's track meets and didn't watch her in the drill team.

All right. She was wrong but she was right. She was wrong he loved Craig more than her, he didn't, they were the apple of his eye, he loved them both. But she was right, he was a better father to Craig.

Because why, because he and Craig had more in common. Drill team. He could not think of anything more boring than standing on the side of a field on a summer day watching a parade. But that was selfish, Debs would've been happier than hell if he'd been there.

There was that tune she used to sing, she would dance it around the house, spinning and closing her eyes, the *Hawaii Five-O* song— ba ba ba ba baaaa-ba, ba ba ba ba ba baaaaa—raising her hands way over her head, dancing her elbows and twirling her hands around. He should have been there because of her, not because of him, that was selfish. Mistake number 1 million—no, 1 million and 5.

So Debbie marched at halftime of every basketball game, and he never went to watch her march. If Craig had been on the basketball team, he would have been there every time.

He signaled for another scotch and then there was the other part. The part that was right there he couldn't think about. Why was he so proud? He was the one that was supposed to be mature.

She was an impulsive person and so was he. They were now-type people, they made snap decisions. Usually he was right, probably in the high 90s being right, and times where he'd stop and debate it, well, that's when he made a wrong decision. Still, how could he have gotten so angry at her? Debs always did what she wanted. They took her hunting and she read that book in the blind the whole time then the ducks came and she jumped up and yelled, "Fly away, ducks! Fly away!" Everyone was angry, how could you be angry at a girl doing that? How could he ever be angry at Debs and that's what made him hate the boy that killed her even more.

Fly away, ducks!

Well, he had been too strict in a lot of ways. He'd followed his dad. His father was a poor country boy and strict as can be, some people said his father was harsh, and Wayne toned it down in most things but not in all of them.

Then he'd missed Fathers Weekend. Her second year at Washington State

Debs had begged him to go. Because he'd gone the first time. But then he didn't go the second because he was in Alaska, well, that was a shitty excuse, Barbara could fly him anywhere in the world for $30. Plenty of time to go hunting, though. Whereas Howie and Dorothy went to Pullman for Susan, they hadn't missed one.

That first Fathers Weekend Debs was proud as a peacock, all glowing and full of herself. He went to the homecoming game then they visited the sorority house. "I'mmm Suzeee Sor-O-Rity," she sang ditzily in the car, making fun of herself, she was the last person you'd ever think would end up in a sorority but she was proud, it was a fancy one, Kappa Alpha Theta, and she held his arm pulling her big strong father through the place introducing him to the girls.

The first person in the family to go to college, damn right she was proud.

I broke my daughter's heart.

How many times had he broken his daughter's heart?

It hurt too much to even think about it. He wrote that angry letter saying she was ungrateful and why wasn't she getting her degree in bartending, and she was a Gardner right back, broke it off without a word, went over to the Peace Corps without even calling or visiting Alaska. "Dad, she says it's a cold war." Yes, and what were they fighting about? Nothing, a check in the mail. And he didn't do a thing to end it, the adult, he'd done nothing.

Nothing nothing nothing nothing nothing nothing nothing nothing nothing nothing nothing

She shouldn't have said those angry words to him, she was emotional, protecting her brother, but she hadn't shown him any respect. Yes, and he should have just let it wash off and, Here's another check.

Alice never made mistakes like that because a woman was born to be a mother, a man has to learn to be a father, and anytime you're learning, you make mistakes. But a woman could see Debbie hurting from what he'd done, and he couldn't see, Alice could see. Oh he could see all that now, he could see everything, and to learn it all now, and not be able to do a thing, oh—and he put it on the room and went up.

.   .   .

Justine drove Alice the thirty minutes back to Tacoma, and they were quiet a lot of the way and Alice thought about why Wayne was so torn up.

He may not have been the best husband but he'd been a good father. The main thing was he loved his children. He did things for the children, he liked being with them and taking them with him everywhere, explaining things. He was strict and he cared about them, sometimes too much. He related more to Craig, but he took Deb into the woods all the time, and he took her to her skating lessons every Saturday.

After the divorce, he didn't see the kids as much. Well, that was par for the course. He was a good father. No he wasn't Father of the Year, but then what father is?

Craig's money hadn't come and he could have gotten kicked out of school, and it upset Debbie so she took up arms for her brother and wrote her dad. It was worse when you put something down in words. It was easier to walk away from someone when it was right there in writing.

Why did she write her father like that? Because she had a special relationship with him. The sun rose and fell with her father, and Deb thought she could tell him anything. She saw an injustice and she hated injustice, and she could tell her father directly about that kind of thing. They took after one another that way, stubbornness, bluntness, forthrightness.

Deb was a little in love with her father, and he was in love with her.

There was a mixup with the Peace Corps about when the plane was going to arrive. Emile was supposed to get there from Honolulu late Saturday night, but he didn't get in till Sunday morning early. He had stuck with the casket at all times, been a pain in the ass to cargo handlers and baggage clerks, till finally at dawn he saw a funeral home director close the gate on a station wagon and he breathed easy.

The guy drove Emile to his hotel in the makeshift hearse and Emile grabbed a few hours sleep, then Craig picked him up and brought him to the little condominium in Garden Court with the small living room and the pass-through to the kitchen.

It was Sunday and the Gardners responded to Emile. He was funny, he was sincere.

He told them about Deb listening to Beethoven's Sixth as the sun rose, about her reading a book a day in her *fale*, about her going to the blowholes on the south coast where the sea came battering in under a coral shelf and drilled holes up through the coral, and her looking down into a hole when a whistling plume of water shot up and knocked her over.

He told about her Chinese one-burner kerosene stove and the treat she liked to fix for him, grilled bread with cheese and celery salt.

Well, Emile had never had celery salt, he was always going to have celery salt from here on. . . .

So in the midst of devastation Emile was a bearded Peace Corps shaman in scratchy missionary pants, a conjurer to bring Deb back to life for a little while anyway.

The Peace Corps had told Emile to advise the family that the remains were not in viewable condition, and Alice had no desire to look, but Craig said he definitely wanted to see her. Wayne hesitated. Barbara thought of saying to him, Go say something to Deb, but she held her tongue, she respected his decision, and in the end just Craig and Emile went to the funeral home and went in one at a time. The casket was on the floor and open. That disturbed Emile, he didn't like to think of Deb lying on the floor. Then when they got back, Wayne said, "How was it?" and Emile said it was pretty bad, and later Craig said that Deb had been cut up, her hands especially were mangled, from fighting Dennis off.

Craig dropped Emile back at the hotel, and he hoped to get a decent night's sleep but the phone rang when it was still very dark. Emile lifted his head, and then he remembered where he was. This wasn't Ngele'ia, an electric clock said it was six in the morning.

"Hello, Emile. My name is Paul Magid—"

Magid was a lawyer in Washington, D.C., for the Peace Corps. He had all sorts of questions about the case, what had happened in the days leading up to Deb's death. Emile sat up and answered them one by one. This was the parallel investigation, and there was caution in the lawyer's voice. He emphasized the importance of giving the parents a "complete but tactful picture." Not just Deb's parents, Dennis's too. No matter that Dennis didn't want him to, Emile should call the Privens in Brooklyn when he got the chance.

A second Peace Corps official was at the condo that morning. Dottie Rayburn was the country desk officer for the South Pacific who had greeted Tonga 16 in San Francisco when Deb shipped out, now she had brought an American flag with her from Washington. She sat with the Gardners and tried to explain the case as best as Peace Corps understood it. They did not have all the information yet. The Tongans had arrested the volunteer, he was in jail. He was a chemistry teacher at another high school. He was outside Deb's house that night, some kind of stalker.

"The trial is going to be in Tonga. The American government will have to pay to defend him."

That was the bombshell. Wayne and Alice and Barbara were silent, upset. Wayne was sitting on a chair next to the couch and motioned with his scotch glass.

"Well, that is kind of weird, isn't it? He did it, why is the American government defending him?"

"It's the law," Dottie said. "We have to defend anyone in the Peace Corps, no matter where they are or why."

Everyone was waiting for what Wayne would say.

"Well, I suppose—" He sighed. "What's the difference, he's guilty, he's in jail, he's going to spend his life in jail."

There was so little real information, Barbara felt like she was in a movie. Something had happened, something else was about to happen. Peace Corps hadn't even gotten it straight what airplane Deb was coming in on. "They lost the body," Alice had said. The absence of information put more weight on Emile. He tried to tell the Gardners about Dennis. Dennis had been his good friend, but in the second year he had started to come apart. People noticed things but no one ever thought he would be violent.

Peace Corps preferred that an official bring a body back because Peace Corps deaths were so laden with geographical and psychological peculiarities. He was walking home through the village at night and somehow fell down the river bank and broke his neck, and no one found him till the morning, we're trying to get more information. . . . It had seemed to others that she was depressed, then she waded into a lake that was known to be filled with crocodiles, her body has not been found. . . . It was instant. He was standing laughing with friends on the dock when a truck backed up to unload things and a pipe slid right off the back and impaled him through the chest. . . . Dinner was on the table, her roommate had made it, and what we've been able to figure out is she turned on the butane heater so she could have her bath. Volunteers had been warned not to operate those heaters without opening a window, well, evidently the flame died out and the house filled with gas, they would have fallen asleep. . . .

Emile was no information officer in a suit, he was a traumatized 26-year-old who had loved Deb, and loved Dennis too. Back in Tonga, Mary had told Emile to be thoughtful about what he told the Gardners. And yes, he was thoughtful, but he wasn't an official, he was a hipster from the Bay Area, an art teacher wearing beads he'd fired out back of his *fale*. The case had already tied the Peace Corps into all sorts of knots, and in Emile it had found an unwitting front man.

On Monday, October 18, four days after the murder, Frank Bevacqua had his thirtieth birthday. Deb had been planning a party for him, and though it seemed crazy to go through with it, it seemed crazier not to. People said she'd want them to have the party. So they did.

Of course Frank thought about all the bad feelings the Tongans now had for Peace Corps, and out of deference to local sensibilities, he had the party at a far western beach on the long curl of land where *pālangis* had first happened on Tonga, in 1643, initiating the era called Contact, and Frank asked a couple of Tongans to stand by and run interference.

The volunteers brought beer and *hopi*, they brought a lot of tapes and then they got as drunk as they had ever been and cranked the music. It wasn't *fakatonga*, but what did they care? There were times when crossing cultures had to go two ways, that was actually part of Peace Corps's mission. And this was the volunteers' culture. They were going to grieve and heal as American youths.

People who'd never had a drink before had one that night, and some kids seemed to lose their minds and began jumping over the bonfire. Whose idea was that? Nobody's, it just happened, though it did occur to Frank, who was as smashed as he had ever been in his life, that this is what that word from college meant, catharsis.

Once the jumping started, someone said their clothes could catch fire, so they took off their clothes. Straitlaced Peace Corps women stripped down to underpants and ran screaming over the sand to hurl themselves through the air. Even Jackson had his clothes off. '*Oiaue* Jackson! *Tapu*-respecting filariasis-studying nurse-aspiring supervol Doug Jackson flew back and forth too low over the flames, burning all the hair off his legs.

As the bonfire died out in Tonga, the news of Deb's death was published for the first time in a newspaper. The Mountain View Funeral Home had prepared an obituary, and the *Tacoma News Tribune* printed it deep inside the newspaper Monday morning.

"Deborah A. Gardner, 23, a Peace Corps volunteer from Steilacoom, died Thursday in Tonga. She had served with the Peace Corps 11 months in the South Pacific."

No more information than that, but "23 years old" and "Peace Corps" was enough to get wrongful-death attorneys calling 2822 Garden Court on Monday, and Craig answered the phone from the couch. "We're not interested," he said each time, and hung up.

Mountain View was a large, lush cemetery outside Tacoma. Narrow avenues wound their way past giant trees and cozy chapels, and the funeral was held outside that afternoon, but not at graveside. There was no graveside. Emile and Deb had talked about death, and Deb had told him that she wanted to be cremated rather than put in the cold ground, she didn't like being cold.

Alice had had her hair done that morning to look nice but she didn't really see anyone else that day, or hear anyone, everything passed her by in a blur, and she thought, Was this going to be her life, a blur? She resolved to go back to work Tuesday, if only to distract herself.

She studied the flowers. The wreath on the casket was signed Alice and Craig, hers of course. There were flowers from her parents Carl and Marie Johnson, from Wayne and Barbara, from Wayne's parents Roy and Agnes, and a lovely wreath of carnations and daisies from the girls on the second floor at the Bon.

A bouquet had come from Peace Corps, too. Mary had seen to that.

SEND A SPRAY WITH THE WORDS QUOTE LOVE FROM PEACE CORPS TONGA UNQUOTE.

Then a dozen red roses and no card. Alice gazed at the roses in the vase. Who had sent the red roses? She had no idea.

The minister was a Presbyterian pastor named Bosteels. He'd asked to see Deb's body that morning, and it disturbed him. He saw a beautiful woman who had been badly beaten up, and he put his hand on Deb's forehead to pray for her. "Lord, this is the symbol of your child."

He had spoken with Craig and Emile about Deb and now said something that Craig had told him: "Deborah Ann Gardner was a free spirit."

Wayne turned to his wife. "What in the heck is a free spirit?" he whispered, and Barbara had to explain it to him. It means someone you can't put in a pigeonhole or a cage, someone like you.

Then the pastor said, "Her father couldn't make it here today." But people said, "Wayne is here, here is Wayne—" and the pastor corrected himself.

"The lord is my shepherd, I shall not want, he maketh me to lie down in green pastures—"

Back at Garden Court there were drinks and light food and more calls from the personal-injury lawyers, and little groups formed at the edge of the company, people trying to get information. From time to time a person said something that caused a commotion, and it wasn't a funeral that gave anyone the sense of an ending. It was the middle of a fairy tale. Grisly events had taken place, no one knew where the tale was going next. Something that was supposed to be calm had happened at the cemetery. Now they were back at the condo and no one was calm and there were tons of questions.

Deb was here, then she was gone. No one knew what exactly had happened to her, and Dottie wasn't much help.

"We don't know much else," the country desk officer said.

Roy Gardner, Wayne's short-tempered father, turned on her. "You know, we can sue."

That stirred Alice from her walking coma. She pulled Roy Gardner into her kitchen.

"Roy, you need to understand something right now, I will never sue. I will never make a cent of money because my daughter was killed."

So Roy backed off. That was Alice. She was formal on the surface but if she decided something, get out of her way, and Wayne echoed his former wife. That was blood money, he wasn't interested.

Wayne's second wife was more detached than Alice or Wayne. Barbara wasn't sure what her place was, but she already imagined sitting at the boy's trial, looking at him, trying to understand why someone would do such a thing to a girl like Deb. Barbara had liked Deb. Deb had the sense of limitlessness that came with growing up in Washington, she was rugged but she was fine, too, getting Barbara to buy her that black leather skirt at Nordstrom's and not caring what her mother or father said. The boy was a chemist from New York, period. Barbara felt that she ought to know a lot more.

"Wayne, do you think you should go to Tonga?"

Wayne looked at her, upset. "Why?"

"I can do something about the fare. I want to go."

"What's the point? He's arrested, he did it, he's guilty, he's in jail. Even if the Peace Corps is going to defend him, what's the difference? Everyone is entitled to a defense. But he did it, he's guilty, it's not necessary."

Then it seemed to Barbara it was too painful an idea for Wayne even to think about, and she didn't push it. He was a conservative, and trusted the government. Alice was the same, and Tonga was out of the question for her. She'd never go there. Well, she would only go there if someone coerced her. No one did.

There were two American flags, the one that had come out on the coffin and the one from Washington, and Alice got one and Wayne got the other. The ashes went into a pottery urn that Craig and Emile brought down past the azalea and rose bushes in Steilacoom to Puget Sound. They distributed the ashes and broke the urn against a rock by the water and threw the pieces into the sound.

That felt right. Deb had walked here, she had sunned herself here, and swum here. She'd lain on the sand on her back alongside her prom date one morning and said, Do you see that one that looks like a dog? and pushed the top of her head against his so their eyes could be closer, so they could try and see the same cloud. No, that one—don't you see? The Pacific was her horizon. She'd pleasured on it and gone out on it, now she joined it.

Emile went back to the Hilton to crash. He'd gotten Craig to take him out for his first real hamburger in two years, but there'd been far too much liquor and sweets and he hadn't slept fifteen hours in four days. He needed to recharge. Tomorrow he was headed out to the East Coast for the first time in his life. To be debriefed, they said, whatever debriefed meant, like some commando on a secret mission.

This time it was a knock. Wayne Gardner was standing outside the door, looking away.

"Oh, hi, Emile! I guess you'll be going now—"

"Yeah, I'm flying to Washington. I mean Washington, D.C. I'm already in Washington!"

"You know something—Debbie and I—we weren't getting along so good for the last year or so."

Emile nodded. "I know."

"She talked about that?"

"She talked about you all the time, Wayne, she was sad about that."

Again the thing heaved up inside Wayne and cracked against his throat, and the worst feelings flooded him. Debs was furious at him. Debs didn't want to see him. Debs didn't like him.

Emile thought about the Dateline dance, the walk out on Yellow Pier.

"She talked about you a few days before she died. She felt so bad about everything, she couldn't wait to get home and make up with you."

"Is that right?"

"She said she was looking forward to that, she loved her dad so much."

Wayne's head bobbed nuttily. "That's good." He turned and sucked a brave breath. "Well, Emile, I guess you'll be going."

"Yeah, I'm so important the government's picking me up at the airport in a limousine."

But what was this now, Wayne Gardner lunging and clobbering him in his arms, a squeeze, a bear hug, this moose-elk from Alaska hugged Emile so hard he couldn't breathe.

17

The Jail

He had one book, the Bible, and never having read it set out to do so now. The cell was rather dark. Dim light fell from a high barred window just under the crumbling gingerbread eaves. He'd lost his glasses, still he bent over the text and read it as he had tried to do everything, thoroughly.

It was Laura who prevailed upon Dennis to eat. Dennis had always disliked Tongan food—Laura brought him hot homemade American food. Laura Boucher was thin, blonde, emotional, thought by some to be flighty, but she had such purity about her, an air of commitment and devotion, and she implored Dennis and even berated him, till at last he ate.

The friends came every day. Dennis's first night in jail, or maybe even before, he had thought to himself, Who will stick with me? and he had made a mental list. The three volunteers from Tonga 16, Laura and Paul Boucher and Paul Zenker, the shy and curly-haired science teacher from Notre Dame who lived out at 'Atele, those he had expected. But not the Bird Lady from his own group. Dennis had not known about her back story in Michigan, how a few people in her town had been willing to talk to her family and how important they had been to her.

In the beginning the police turned the friends away. "This is a jail, it is not made for receiving visitors," the officers said defiantly and fearfully. "Jailed prisoners cannot receive visitors." Still they came, and as it was a special case and the police were not sure how to deal with *pālangis*, they were accommodated.

Dennis's former director came in from Samoa to see him, the psychologist Dick Cahoon, and along with Cahoon a small psychiatrist with an eastern European accent and wearing an Aloha shirt. He met with Dennis over two days. Peace Corps offered a mixed message about his presence. It said that he was there to help all volunteers who were traumatized by the events. Or that he was stabilizing a volunteer who'd tried to commit suicide. Maybe also he was preparing a psychiatric defense against criminal charges, but that was not to be said aloud.

Paul Boucher met with the psychiatrist for three hours, Paul Zenker too, and later the Pauls shared some of the insights with the Bird Lady. The psychiatrist felt that Dennis had been deteriorating mentally for a while. The signs had been there but people had not picked up on them.

The crown solicitor had announced that a preliminary hearing would be held in magistrate's court two weeks after the murder to determine whether Dennis was to be held for trial, and the friends sat eating with the prisoner at the long stained desk in the charge office and talked over his strategy in the case. The attentions seemed to disturb the police. They felt that the *pālangis* ought to be praying for Dennis, not chatting with him. "I almost hate going down there," the Bird Lady told her parents. "I never know who I'll meet—friendly or unfriendly policemen, maybe or probably not see Dennis, get to leave a message or not, be chased out or not. It's always a tense ride down."

The police compound stood near Dennis's old *fale* in Longolongo, encircled by a high fence with barbed wire strung along the top. When an official vehicle came through the gate a policeman stepped from the shadow of the guardhouse and snapped to attention, his rusty rifle leaning against the whitewashed stucco wall.

A few steps west of the headquarters building, the police minister's house was a low *pālangi* house surrounded by banana trees and the red fanciful shrub called *si* that looks like a feather duster, from which the Hawaiians make hula skirts. There the minister was on his little patio, smoking a cigar and drinking coffee, paging through a book of Shakespeare, or Freud. 'Akau'ola had spent time in the West, he had Western tastes. But he was a severe character, with unpredictable moods. His small seedlike eyes seemed smaller on account of the glasses he wore, and his great height and slablike shoulders. He spoke perfect English and quoted Browning.

> *The only fault's with time;*
> *All men become good creatures: but so slow!*

The minister was vexed. From the time late Thursday night that his officers had told him of Deb's murder, he had perceived that a bright light would at some time be shone on the matter and he'd instructed his detectives to work twenty-four hours a day. But none of them was deeply experienced in investigation. Cases in Tonga were open and shut, criminals confessed. This man said nothing. And from the start 'Akau'ola's men bumbled.

They had the sense to lock up Deb's house, but when the minister got there in daylight, he walked around it shaking his head.

"These bicycle tire prints—to whom do they belong?"

"We will find out," said a constable.

'Akau'ola waved his hand, dismissing the man. "Now it is too late, don't you see? You have not properly secured the scene, the entire scene should have been secured at once."

He held that and other errors against the man whom Mary saw in her vision, Maeakafa, the superintendent of criminal investigation for the city, and Maeakafa struggled. He had little knowledge of murder, he had to school himself in his notebook. "Charge for murder. Murder is like an egg. ie Murder is an yolk. The white substance is manslaughter." He did not undertake a thorough search of Dennis's house until 5 P.M. Friday, by which time a horde of Peace Corps people had already been through it, in the parallel *pālangi* investigation, though he did manage to pick up a left flip-flop there—"coloured red, the holding leather coloured green"—and determine that it matched the right flip-flop he had found inside Deb's door.

There were not many other clues. Two forensics officers spent the day in

Deborah's *fale* with black powder everywhere, still they could lift no fingerprints at all. One would have thought that the knife would yield prints—but the Seahorse was hopeless. Too many awestruck officers had handled it, marveling at the serrated blade, showing it off to the silver-haired chargé d'affaires.

As for photography, alas, two constables were at odds over the assignment, and Faiva got the job and having less experience than Nauto failed to check whether the magazines for the camera were full. Of course it turned out the magazines were empty, so all the images of the crime scene were lost. Nauto was then sent out to take fresh pictures, but by then Emile and Brombach had cleaned the hut. There were only two or three blood stains to be found. Nauto had no chalk either, with which to indicate the position of the body. So he laid eight size-D batteries on the floor in a cruciform aimed at the door, and photographed them.

Then the minister came out from behind his desk to yell at the detectives. He said that they had failed in basic investigative practices, that they were uncoordinated.

The minister did not let on, but he was also pleased by some of the police work. The detectives had been careful in their custody of physical evidence. The flip-flops. The eyeglasses. The syringe, with a small case of needles. A short length of metal pipe. The bottle of mysterious powder. And the jar containing a bottle containing the *tapu* fluid.

They had done a good job of interviewing Tongan witnesses, and two or three days after the murder, Faka'ilo had been able to question the accused, the interrogation that Mary saw as a violation of his human rights. The police had let Dennis out of his cell to sit in the charge room, and the top detective had asked questions that Dennis had by and large refused to answer. But in a couple of brief utterances he had surely given himself away.

"Do you wear glasses?"

"Yes."

"Where are your glasses?"

"I don't remember where I left them or when I wore them last."

"Look at these glasses—"

"I don't want to say anything about it."

"Look at this knife."

"I have nothing to say."

"Do you know Mr. Emile Hons?"

"Yes."

"Who is he?"

"He's a Peace Corps volunteer."

"Have you ever visited his house?"

"Yes."

"Is there another house located together with Mr. Emile Hons's house?"

"Yes."

"Whose house is it?"

"I do not know."

"Have you been to that house?"

"I have nothing to say."

"Do you know Miss Deborah Gardner?"

"No."

"Was she a friend of yours?"

"I have nothing to say."

So he had lied twice about Deb. That was damning.

Two days after the police interrogation, a Land Cruiser swept through the police compound gate, and Haini Tonga, the commander who was second to 'Akau'ola, heard a soft knock on his office door and looked up to see Mary George. Unlike his boss, Haini was a friendly sort. He told Mary to come in, and she took his hand in her bone-crushing handshake before sitting down.

"The Peace Corps would very much appreciate it if you would drop the charges against Dennis," she said.

Haini pushed back in his chair and smiled at Mary in a mild and infuriating Tongan manner, showing nothing, waiting the other person out, waiting for her to commit, and now Mary was compelled to repeat her request.

"I am asking you to withdraw the charges against Dennis."

Haini smiled. "But the police have no power to do that even if I wanted. The case is a formal proceeding. It will come before the magistrate. If you have evidence that the police have arrested the wrong man, you must bring it to the attention of the magistrate."

Which Haini thought was the end of it, till he went to a cocktail party at the Dateline a day or two later, and Mary came up and reached for his arm in her touchy way and made the same request again.

This time Haini was blunt.

"Don't you pity the victim?" he said.

"Yes, but now we must devote our attention to the accused. He is living and the girl has died."

"Well, I am in sympathy with the victim and her family," Haini said. "I want to help the family of Debbie to drown their sorrow."

Mary regarded him with her penetrating sharpness. "We now must concern ourselves with the living volunteer."

Mary's advocacy for Dennis was not only personal. One of her chief responsibilities as a country director was to protect her program. The Peace Corps manual was emphatic on this point. If a country director judged her program to be at risk, she was granted wide latitude in sacrificing individual volunteers. A volunteer who generated suspicion about drug or political activities could be sent home virtually without a hearing. Now the Tonga program was at risk—who in the tiny country could focus on the task at hand with Dennis in jail? By trying to get him out of the country, in a hurry, Mary could be said to be serving the program.

A perception had formed in the *pālangi* community that the Tongans wanted nothing so much as to get rid of the case and that with the proper application of pressure they might be caused to cough Dennis out like a chunk of meat from the windpipe.

The case did not really involve Tongans, it was *pālangi* on *pālangi*. And though the treaty under which Peace Corps volunteers lived in Tonga did not give them diplomatic immunity, still a vigorous prosecution could only damage Tongan's relationship to the United States. Tonga was broke, it had lately played the Russians off against the free world so that New Zealand would come through for runway lights at the airport. Why should it do anything to alienate the Americans? "The King and Tupouto'a have expressed sentiments about helping in whatever way they can to get Dennis out of here," Mike Braisted wrote in his journal. "Nevertheless it will probably take at least a month. A very rough time for Dennis."

Crown Prince Tupouto'a was the King's tall dandyish son, and the first secretary for foreign affairs. He had offered assurances to the chargé d'affaires and to Mary George. He did not see the crime as premeditated, he believed that Dennis was "deranged by passion." The chargé then told the Ambassador that the Tongans would just as well see the case disappear.

Yet Tupouto'a had said something else that the chargé did not emphasize in his reports: Tonga had an independent legal process that must now unfold.

This process was no charade. Tonga had had the rule of law for 101 years; in 1875, partly at the urging of that rogue missionary, Shirley Baker, who had become prime minister under His Majesty's great-great-grandfather, King George I, the Kingdom had adopted a constitution. *Lao 'o Tonga*, or law of Tonga, was based on English common law and made up three squat black-bound books that were frequently amended by the Tongan Parliament and that 'Akau'ola kept on his bookshelf.

One of the King's major initiatives had been to fill his government with professionals, and many of these professionals were in law enforcement, three of them British subjects whose services were funded by the British Overseas Development Administration. The *piusne* judge (pronounced "puny," a lower-ranking judge), the crown prosecutor, and an adviser to 'Akau'ola were all seasoned *pālangis*. The crown prosecutor in particular was a coldly professional, wire-rimmed young woman called Christine Oldroyd who already had years of experience in English criminal courts. Oldroyd's boss, Crown Solicitor Tevita Tupou, was the leading lawyer in Tonga's young generation and had spent some years practicing in England, too. These officials came to the case with deadly seriousness. All would say that a 29-year-old prince's opinion was of little importance to them.

Of course, the murder had a political dimension. Under the law of Tonga, if Dennis was sentenced to hang, he could appeal that sentence to the Privy Council, the King sitting with his Cabinet, and at that time, the American government could issue appeals to the prime minister, Prince Tu'ipelehake, the King's brother, and to his first secretary, the Crown Prince, as well. The American press and politicians might speak out about the case, and how likely was it that Tonga would then hang an American?

The political dimensions, however, were not just international. They were also Tongan. Everyone in Nuku'alofa was watching, and understood the case as a test of their modern system.

No one in the United States knew about the murder, which was fine with the Peace Corps, but meantime it was public in Tonga in a thousand ways. A village, a crowd of nurses, and two doctors at the country's largest hospital had tried to save Deb. Students at the country's premier educational institution whom Deb had taught parasitology and how to sew a teddy bear had wailed through the night at the morgue and gathered with hundreds of others, including teachers, preachers, and administrators, to see her off at the airport. The Ministry of Education and Works was starting a scholarship fund in her memory, while the two top schools in the country were scrambling to replace the science teachers they had lost to the case.

News of Deb's murder had reached the most remote islands, thanks to A3Z, and one week after her death (and two days after the obit in Tacoma) the *Chronicle* published the first printed news of the murder anywhere in the world on its front page. The article stated that her body had been sent back "to her parents," and an accompanying photograph showed "Miss Deborah Gardner" in a tender mood, her head tilted. Throughout the life of the case, Frank Bevacqua's picture was the only photograph that would ever be published.

There could be no question that Tongans deferred to the Crown Prince and his father. The line was ancient and godly, the commoner's relationship to royalty was governed by the strictest *tapus*. When Tupouto'a came into a room, people fell out of their chairs to lower their heads before him. When he went by on the street, cyclists and drivers alike pulled to the side of the road or were ordered by police to do so, and people bent their shoulders.

Yet a great number of Tongan officials now had some role in the case, and none wanted it to disappear. Several of them had known Deb, several saw it as their duty to see justice done. From *Chronicle* editor Paua Manu'atu to police secretary 'Eleni 'Aho to education and works minister Langi Kavaliku to A3Z director Tavake Fusimalohi to magistrate Pahlavilala Tapueluelu, to Palace Secretary 'Epeli Hau'ofa—professional Tongans had heard talk about what they called "the directoress's agenda," and it affronted them.

'Akau'ola's original name was George Faletau, but he was from a chiefly line. Thus the title, 'Akau'ola, the ancient title of one of the *matāpule*, or talking chiefs for the King. The 'Akau'ola accompanied the King at sea, and even now 'Akau'ola sometimes appeared alongside His Majesty with dried seaweed draped over his shoulders, to perform his *fatongia*, or traditional duty.

Duty had formed 'Akau'ola. After his second year at the University of Auckland, His Majesty had ordered him to withdraw to become a policeman, and though he had wanted to stay at university and read literature, 'Akau'ola had obeyed His Majesty and gone to the police academy. Then he had served in the police force in Auckland and fallen in love with a red-haired policewoman named June, but his own family had directed him not to marry her, and he had obeyed.

As a boy, his father had been harsh to him, marching him naked around the yard after he had flirted with a girl, shaving his head, making him pick up dogshit with his bare hands, all in the name of discipline. His mother was the free spirit, but his mother was gone. "Your mother is no longer with us, I've told her to leave," the father announced to his sons one day.

The eldest grew into a hard man, and a fair one. His policemen said that he had "no pony"—no favorites on the force. But in those areas of life to which duty provided no railing, 'Akau'ola was neurotic. He never married, he was married to the job. He had girlfriends, but he lacked the feminine hand in his life. "Talking to 'Akau'ola is like talking to a man behind a veil," said one *pālangi* diplomat. The King picked out a noblewoman for 'Akau'ola to marry, but the woman had eloped rather than take the veil with this man who wore one himself, and now it was said that a strong-willed female cousin within his own household ruled 'Akau'ola, and 'Akau'ola was ridiculed for that.

Tongans said that 'Akau'ola had never married because he did not want to bring upon his descendants the *mala'ia*, or negative karma, that was his for having killed a man. 'Akau'ola had performed the last execution in Tonga five years before. Personally he did not believe in hanging, it went against his religion and custom. But he had been compelled to do it because no one else wanted to, and he saw the job as the police minister's ultimate responsibility.

He agonized, he had bad dreams about serving on the gallows. Still he did it, and later, in church, 'Akau'ola held up his hands. "How do I confess to my maker that these hands are now bloodstained? Through the blood of Christ, these hands will be cleansed."

The minister had seen the nightdress stiff with Deb Gardner's blood, and he had asked his men what the girl was like. The detectives all said the same thing, She was a good girl, she was well-liked by her neighbors. She was modest and had a big smile. She had baked an orange cake and other treats for the Pasas, in whose truck she had made the last journey of her life.

Deb's character was being smeared in some quarters already, and it was whispered that she had not been wearing underpants when she died. What did 'Akau'ola care about that? He pitied her. She had come out across the sea only to help. She had put tapa on the walls of her house, then been savagely attacked there and had died in the arms of strangers. The girl's parents had not appeared. Meantime, the girl's own organization seemed to obstruct the investigation.

The detectives were experiencing difficulty with *pālangi* witnesses. The police were trying to establish the social background to the murder, but either they were unable to ask *pālangis* about personal matters or the *pālangis* were holding out. One way or another, they were getting very little information.

Rumors swirled about the McMaths. What had happened at their house? The morning after the murder, neighbors had seen fear in the McMaths' eyes as John said firmly to Marie that she must lock herself inside with the boys as he rushed off to Gay's house. Maybe Dennis had threatened to kill the McMaths' dog Brutus: that was one rumor.

But when the police asked John McMath to analyze the chemicals, as John was a science teacher who was familiar with all the chemicals in the Tupou High supply room, John refused. Analyzing the chemicals was a job for a forensic officer, not a deputy principal, he said.

John sensed that the police were in over their heads, panicky. Many *pālangis* were fearful of what the Tongans would do, and when Marie wrote to her parents way off in Australia and told them about the murder, she cautioned them to silence.

"I suggest that you now sit down if you are not already doing so. Also please do not mention this to anyone but the Lord. Something terrible happened in Nuku'alofa last Thursday night, which has deeply affected us and all volunteers, in fact, the whole island. Dennis did his washing here on Thursday afternoon; and some time that evening something terrible took his mind to do this thing. . . .

"Please do not press for details and do not comment to anyone please."

The Pauls held murmuring conversations with Dennis about evidence as policemen moved about them in the charge office and now and then seemed to try to listen in. The Pauls were encouraged. There were many police foul-ups in the case, the evidence was weak. It looked as if Dennis might be released to be institutionalized in the United States.

"I hope so—& I just wish it could be done quickly," the Bird Lady wrote home. "Then just the problem of getting Dennis help."

The Bird Lady let the Pauls talk to Dennis about the case. She spoke with Dennis about day-to-day doings, what was happening at Tupou High. She told him about the two volunteers who had taken over his classes, and she brought papers to Dennis to sign, recommendations for his sixth-form students. Or she consulted Dennis on lesson plans. The Bird Lady wanted to talk to her class about a birth control vaccine that was being tested in India. Dennis knew about it. He explained the way that it was supposed to produce antibodies to certain reproductive hormones.

"Take care of yourself," Dennis called out in his meek, high voice as he was led back into the cell, and the Bird Lady crossed the street to the market to look for shells and then stood by the wall crying. She tried to tell herself that Dennis was actually innocent. If he had gotten really upset he might have slapped Debbie Gardner around. But if Dennis had cold-bloodedly decided to kill somebody, you would never know it was a murder, much less who did it.

A lady who sold baskets came up and offered the Bird Lady condolences. *"Faka'ofa a' Tepi."* The Bird Lady didn't correct her. Then the woman expressed her thankfulness that it wasn't a Tongan who had done it, and the Bird Lady didn't know what to say. She hadn't known Deb well at all, didn't like to think about her. Paul Zenker said it offended him when people rationalized Deb's death, but the Bird Lady recognized that she was rationalizing it in her own way. "The murder is incomprehensible," she declared. Defining something as incomprehensible seemed to relieve her of any obligation to try to understand it, and the truth was she didn't really want to understand this.

Dennis's life was ruined, the only question was how. No one would be helped if Dennis was hanged. Neither would it help him if he was imprisoned at Hu'atolitoli, the inland prison farm.

"Yes, Dennis has to face what he's done & learn, but that will take someone far wiser than me to guide him," she told her parents.

And Gay Roberts wrote to an Australian volunteer who had worked with Dennis, "Although it must have been a terrible way for Debbie to die she is now dead and there is nothing that can be done for her. My sympathy lies with Dennis who obviously needs psychological help and this he will not get here."

Gay visited too. Dennis did not look at her, but smiled, saying he wanted to shave his beard but that it seemed unlikely the police would let him have a razor. John McMath visited, and so did other colleagues. Mesui Saafi, a Tongan teacher, appeared with food from a feast, and the police brought Dennis out and Mesui said a prayer for him and when he unwrapped the cake, Dennis gobbled it down. Frank Bevacqua went to the jail along with a couple of other volunteers and only stared at Dennis, said nothing, and it chilled Frank that all Dennis talked about was Dennis, his own future.

Altogether Dennis became such a distraction that the minister directed that the prisoner be removed to the prison in the middle of the island.

Just four days before the preliminary hearing an open-bed police truck carried him away. The truck moved slowly down the King's Road in a form of public humiliation. Dennis sat on a bench in the open air with four or five officers on either side of him leaning against wooden stanchions. His face was beet red, his eyes downcast, and already he was on another hunger strike, demanding his return to the jail. His friends were all that kept him alive, he said. "I have nothing else to live for, but this contact," he told them.

The minister had little pity for Dennis. He felt that the murder was not a crime of passion, it had been planned. Weapons elevated a crime. If a person used his feet or hands to kill someone else, that was a crime of passion, but if he used a bush knife, that suggested premeditation, and he would hang.

One night the minister saw the Crown Prince and made this point rather firmly to Tupouto'a.

"It is a simple case. The accused was a chemistry teacher. He brought a pipe, syringe, needles, chemicals, and a serrated knife to the young lady's house, then stabbed her twenty-two times. There is no question what should happen to him.

"If necessary I will hang him myself."

The prison was called Hu'atolitoli. It was twelve miles from town, and Dennis was placed in maximum security. Even then Laura managed to get him a blanket, while Mary beseeched the police to bring him back to the city. Dennis's friends were cynical, saying that Mary was just trying to save her skin, or she felt guilty for not having done something about Dennis when she could have.

But whatever her motivation, the effect was good. Two days after they'd paraded him out of town, the police brought Dennis back, and Laura brought dinner.

The shadow of the noose filled common Tongans with compassion. They expected that Dennis would die, but also seemed to pull for him against the police. Tupou High School students came to the jail to pray, or pretend to anyway.

"Why did you do it?" said one.

"Time's up," Dennis said, ending the meeting.

One day, there was such a big group of students from Tupou High that the police brought Dennis out barefoot to the big banyan tree. The tree's roots formed knees that you could sit on, and Dennis sat with his back against the tree. Two or three policemen stood watching, and the boys in their skirts and mats sang a hymn, then one by one they went up to Dennis and shook his hand. A teacher had status, even a teacher in fetters, and a couple of boys hugged him.

"Mr. Priven, what happened?"

Dennis looked down. "It's a long story—"

The school chaplain had accompanied the boys. He opened the Bible to read, but he had only gone a verse or two when the writing got in his way and he set the Bible down on a mat and spoke haltingly as if he were speaking directly to Dennis. The chaplain was a sweet-natured man with a low forehead and a strong voice.

"The second son request from his father money. Then he expend all he received with his friends. For what he wanted. Drinking. Gambling. Dancing. And when it was empty, his pocket, his heart came back. He remembered his father.

"He came begging someone to work. But no one accepted him until a farm accept him. But only to keep his pigs. And after he fed the pigs, some of the pieces were left, and he used it for his meal. Then his heart came to repent.

"'I am so hungry here. At my father's house, the leftover was to the servant. So maybe I can go back and ask his permission, to come work to him as a servant. And not a son anymore.

"But when he came back his father knew he was coming. And RAN to meet him. ACCEPTED him! HUGGED him, and ORDER to his people—"

Now the spirit of the tale overcame him, and Vili Vailea's voice rang in a hoarse bark.

"BRING!! BRING the new sandal! Because he came barefoot. BRING new clothes! And MAKE a SUPPER for a thanksgiving feast. Because this is the son who was lost. Is found! Was dead. Is alive!"

The long-limbed banyan tree made a kind of church, and the barefoot second son sat in shadow, crying. The minister touched him on the shoulder.

"Dennis, we know you are guilty, but God has forgiven you whatever you have done. If you can repent and have faith in God, you will be a new man."

To which Dennis said nothing while Vailea kept at it in the way of all ministers.

"I emphasize you, try go ask forgiveness from the girl's family. I know it is hard for you to meet the parents and the family. But I am emphasizing, do it if you can. To have God's love. God only can reform a new life."

The boys sang another hymn, and Vailea murmured, "May God bless you, Dennis, also the relatives of Debbie," and the police brought the prisoner back to cell 1.

18

Extraterritoriality

One of the most powerful books ever written was *The Ugly American*. It changed its society.

The Ugly American came out in 1958 and was set in a made-up place called Sarkhan that could have been Vietnam. Almost all the Americans in the novel were having a good time at the Embassy or hanging out with one another, but when it came to Sarkhan, they were hopeless. They didn't speak the language, didn't like the food, and were basically getting away with murder, while their Russian counterparts were living close to the ground and winning hearts and minds.

"There are built-in servants!" a female officer wrote home. "They do the cooking, cleaning, laundry—everything. Oh how they baby us! . . . There are only about a thousand Americans here, and we stick together. That means that we girls get asked to everything. I've been to the ambassador's parties several times . . ."

No, it wasn't subtle, but then the authors, Eugene Burdick and William J. Lederer, were two overseas hands on an urgent errand, to inform their countrymen that if the foreign service didn't straighten up, the United States would lose the cold war. And they were successful. *The Ugly American* was a giant best-seller. Everyone in Washington read it and cited it, including Senator John Kennedy.

The Ugly American was actually the book's hero, Homer Atkins, a humble volunteer living with his wife in a village and building a pump with bicycle parts to move water around rice paddies. So the novel laid the ground for the Peace Corps.

The idea had been around for a while by then, Senator Hubert Humphrey of Minnesota and Congressman Henry S. Reuss of Wisconsin had long promoted the possibility of an international volunteer agency run by the government, but it took a clever novel to give the idea political life.

That and a presidential race.

In mid-October 1960, Kennedy and Nixon held a fierce debate over foreign policy in which the young senator said that the best way to keep small countries from going communist was to increase American prestige, and then he flew from New York to the Midwest to make an unannounced 2 A.M. stop at the University of Michigan. The university lifted its curfew for female students, the Student Union was jammed, and in the dead of night Kennedy reached out for a new idea.

"How many of you are willing to spend ten years in Africa or Latin America or Asia working for the United States and working for freedom? How many of you who are going to be doctors are willing to spend your days in Ghana? On your willingness to do that, I think, will depend the answer to whether we as a free society can compete."

The students met his challenge with cheers and tears, and the senator told his people to polish the idea and days later in San Francisco Kennedy used the name "Peace Corps" for the first time, and some claim that it gave him the election. But it was that night in Ann Arbor that was to live on in Peace Corps mythology, commemorated in posters at headquarters and a plaque at the university: October 14, exactly sixteen years before Deb Gardner's October 14, which was also near the end of a close presidential campaign.

The creation of Peace Corps was "a towering task," as the framer Warren W. Wiggins termed it, and a central tenet of the undertaking was that volunteers would not have diplomatic immunity. Diplomatic immunity was a demonstration of arrogance. Foreign service officers did stupid things, they were investigated, and then the embassy invoked diplomatic immunity and whisked the malefactors out of the country. Diplomatic immunity shared roots with the colonial idea of extraterritoriality, a powerful country's belief that its jurisdiction extended to foreign lands.

The Peace Corps wouldn't make that mistake. The volunteers were going to be humble as Homer Atkins, the treaties with host countries would state that volunteers were subject to host country laws. Sargent Shriver made the point when he was selling Peace Corps to Congress, and Wiggins, who would always count *The Ugly American* as one of the most important books he'd ever read, said the same thing in countless speeches defining Peace Corps to business leaders, intellectuals, national associations of psychologists and nurses.

"The Volunteer works within their system for them.

"He does not have a PX or a commissary.

"He does not have a private automobile.

"He lives the way they live and under their laws.

"If he gets into trouble, he is judged by the host country judges and if he goes to jail he goes to the host country jail."

Every volunteer got the Peace Corps handbook, which expressed the idea in almost religious terms. Volunteers had a powerful role to play in the world. They were eliminating hunger, disease, ignorance, injustice, and hopelessness. And along with that power came RESPONSIBILITY. A volunteer was responsible for his own behavior. A volunteer "can expect no special treatment." This code was so absolute that if a volunteer got into trouble, the agency wasn't going to spend any money to defend him, though it might help him find a lawyer.

Then the first volunteers went out to Nigeria, the Philippines, the Caribbean, and the founders braced for scandal. Shriver had assured Congress that the volunteers weren't teenagers, they weren't even "young people." Still, something was going to happen. It was not possible to send thousands of people around the world without some dreadful personal situation arising. A brilliant escapist twisted communist oversexed homosexual homicidal drunken volunteer was going to lose his or her bearings in a river bend that only Joseph Conrad had imagined and make headlines that would bring shame to the program.

But when the first scandal came, Peace Corps breathed a huge sigh of relief.

On October 14, 1961, a volunteer in Ibadan, Nigeria, dropped a postcard she'd been intending to mail, a "Host Country National" found it, and the next thing it was in the newspapers: "We were not prepared for the squalor and absolutely primitive living conditions rampant both in the city and in the bush.

Everyone except us lives in the street, cooks in the street, and even goes to the bathroom in the street."

Anti-American demonstrators filled the streets of Lagos, and many at headquarters feared that Nigeria would bounce Peace Corps, but in the end it was enough that Margery Michelmore go home. A crowd of reporters met her at Idlewild airport in New York, and she held a press conference to express her remorse, and the public got an image of the bad volunteer: a striking and articulate Smith College grad who didn't hold her papers tightly enough under her arm.

Then five years went by, and Peace Corps got the scandal it had been waiting for.

One Sunday in March 1966, peasants living outside a village in northern Tanzania heard cries and ran from their shambas to see a white man crouched over a white woman on a nearby rock feature. The man appeared to be subduing her. He ran to a bicycle, and peasants swarmed up around him, stopped him from going anywhere. The woman soon died, the man was charged with her murder.

Peverley Dennett, 23, of Connecticut, had met William Kinsey, 24, of North Carolina, during Peace Corps training, then married him. A few months later they had gone off on a picnic and endeavored to climb the rock. Kinsey heard the clink of a beer bottle against stone and turned, he told the police, to see his wife slipping off the side of the rock and falling on her head. The injury caused her to flail about, and he needed to restrain her before going to get help.

Bill Kinsey was an outsize character, gregarious and bespectacled, a Tidewater veterinarian's son inclined to fill his diary with dark quotes from novels he was reading. "'You want to know why?' she yelled. 'It is because nobody wants to know why, it is because nobody wants to know anything. Everybody hates everybody but nobody knows why anybody gets shot.'" Writers came from all over East Africa to cover the case, which became a running story on the front page of Tanzanian newspapers.

Peace Corps said it would find Kinsey a lawyer, but he would have to bear the cost. That was Peace Corps code—personal responsibility.

And just like that the code crumbled. Kinsey's volunteer friends said they expected more from the agency, and the North Carolina congressional delegation said that the agency had abandoned the volunteer, a white boy imprisoned in Africa, and President Lyndon Johnson even stepped in on the young man's side. So Peace Corps caved. It would pay for Kinsey's defense, but it said that the Peace Corps Act would have to be amended to permit such a thing.

Peace Corps paid for an international team of lawyers for Kinsey, and even paid for Peverley's mother to fly to Tanzania to attend the trial. She was on Bill's side. She testified that Peverley's letters home had described a happy marriage.

Kinsey was acquitted. The judge said the case had not been proven beyond a reasonable doubt, though he also said that the evidence was sufficiently inconclusive that Bill Kinsey would have to "carry with him the suspicion that he may have been responsible for his wife's death."

Kinsey declared that he wanted to stay on in Tanzania to do TB work, his future was as an Africanist. Peace Corps said that was crazy and brought him back to Washington (to edit the Peace Corps magazine!), and meanwhile Peace Corps lawyers struggled over the change in philosophy.

Would paying for a volunteer's defense upset relations with the host government?

What if the Peace Corps had evidence touching on the volunteer's guilt— was it obliged to produce it?

What if a Peace Corps volunteer was the victim of the crime? Should Peace Corps fund the prosecution as well as the defense?

All these questions would be relevant in Tonga. The case upset the host government. The Peace Corps had evidence touching on the volunteer's guilt. A volunteer was the crime victim, the prosecution was just about broke. But these questions were not asked in 1976. The culture had changed. No one was reading *The Ugly American* anymore. Peace Corps was another bureaucracy, fighting for its existence on Capitol Hill.

From the moment that Dennis came dazed into the police station late Thursday, Mary's first question to Washington was whether he was not covered by some grant of diplomatic immunity to be tried, or maybe not tried, back in the United States.

PLEASE ADVISE INSTANTLY ON WHAT VOLUNTEER'S RIGHTS ARE
FOR PROTECTION UNDER US LAW IN SUCH A CASE.

Washington corrected Mary emphatically.

PCV HAS NO RPT NO DIPLOMATIC IMMUNITY FROM TONGAN
LAW. PCV IS NOT RPT NOT SUBJECT TO PROSECUTION IN THE
UNITED STATES ON ANY CHARGES ARISING OUT OF THIS
SITUATION.

Once Dennis had an attorney, Mary was to stay out of the case. Peace Corps
lawyers would show up to monitor things. The American government's role was
merely to ensure a fair process.

That was so much wishful thinking. The Peace Corps was the sole American
presence in a tiny country, volunteers were embedded in the life of the capital.
They taught in the schools and helped put out the newspaper. The case would
never be about one volunteer. The murder charge threatened the existence of the
program, and Mary was going to preserve her program.

The chargé d'affaires was of like mind. He had been finding his way around
out-of-the-way places before *The Ugly American*, and after it too, and though he
had spent only a few hours in Nuku'alofa, his primary concern was political.
Flanegin wanted to be sure that the case did nothing to damage the relationship
between the United States and Tonga. If Dennis ended up serving a long prison
term, or was hanged, he said, that could only cause awkwardness between friends.

PRINCE REPEATEDLY SAID THAT PEACE CORPS HELD IN GREAT
AFFECTION BY MOST TONGANS AND THAT SYMPATHY RATHER
THAN ANGER IS PREVAILING SENTIMENT REGARDING PRIVEN.
JUDGING FROM WHAT I WAS ABLE TO LEARN DURING ONE-DAY
TRIP, PEACE CORPS OPERATIONS IN TONGA NOT IN JEOPARDY
DESPITE IMPACT OF MURDER ON SMALL, REMOTE COMMUNITY.

The next high American official to come to the Kingdom shared this concern. Jack Andrews, the blocky and buttoned-up NANEAP director at Peace Corps, was the bedrock of postwar society. A World War II veteran and Harvard Business School graduate, he had managed a major department store chain in the Midwest and rescued a Vermont college from financial ruin. He'd lately broken off his tour of outer NANEAP to try and settle things down, and he had no truck with the Ugly American.

There was no point in Peace Corps maintaining neutrality. In Jack's view, the case could only embarrass the American government and hurt the Peace Corps.

Jack and Mary met with Dennis's new attorney, a Tongan from New Zealand, and told him that they didn't want him to bring up drug use by Americans in the effort to defend Dennis. If drugs came up, the program could be thrown out of the country. The attorney said cordially that no one was going to tie his hands, but that he didn't think that drugs had anything to do with the case anyway.

Then Mary took Jack to the market, where they saw the Bird Lady looking for shells. Mary ran up to offer the Bird Lady her fullest sympathy. "I want you to know that everything that can be done for Dennis is being done," Mary said.

The Bird Lady smiled wanly. While inside, she said, someday when she gushes over me I shall scream!

The Bird Lady also ran into Laura, who informed her that she was on a list of volunteers the new lawyer wanted to talk to. He seems trustworthy, Laura said, but she warned the Bird Lady to be careful about what she told him.

"There may be some things Dennis said that should not be passed on."

The Bird Lady understood what that meant. Five nights after the murder, and following Dennis's first day with the psychiatrist, his friends had sat around him in the charge room as he ate his dinner slowly and pursued a discussion of the case. The Bird Lady had been at the periphery of the conversation and had heard only fragments of the talk, and considered herself fortunate in that regard. Still she had gained some inkling into Dennis's plans Thursday night.

"Much cryptic talk of things in his house & things that should be investigated," the Bird Lady wrote home. "If this lawyer is good & can be trusted, & there is a law or precedent that matters between lawyer & client are not to be disclosed, then I feel we should be as open as possible with this lawyer. However, I suppose it has to be Dennis's decision rather than ours."

The Bird Lady had ceded Dennis a surprising level of control. But then she was hardly alone. A number of Peace Corps people had now allied their interests with those of a man they thought to be insane, and a killer.

When Mary shared her vision that Dennis was falsely accused with one of the staff members, Pila Mateialona, the driver who had heard Deb's last words, he got angry at her. "Don't you ever go outside and mention to anyone that a Tongan did this," he said stoutly. But Mary was not so easily contained. She passed that belief along to other *pālangis* and caused camps to form among expats over the question of whether Dennis had done it. And she passed it to Tacoma, too.

Dear Mrs. Gardner:

Since the terrible happening of last Thursday, I have been trying to find words to tell you and Mr. Gardner how completely I share the grief that is yours—

The condolence letter started out conventionally enough, though it soon departed the realm of etiquette and turned to Mary's preoccupation, the living volunteer.

Emile will have told you all we knew of the event when he left, so I shall not go over those details again. The police are holding Dennis, but it still must be proved that Dennis is responsible for the crime. And there is a possibility that he is not the one guilty of causing Debbie's death. I tell you this because I hope your sorrow will be untinged by any hatred toward him. Whether or not he is guilty, he is a young man in great need of our help and our prayers and, as you will understand, it is my duty as well as my desire to give him all the help we can.

*I can write no more now, but please know that all of us here are shar-
ing your shock and your sorrow.*

Alice felt sympathy for Dennis's parents. She wondered what sort of ordeal
they were experiencing. But she did not question his guilt for a moment—and she
was glad that he was in jail. She did not want to pray for him, and she had little
understanding of Peace Corps policy, didn't know that it was not repeat not
Mary's duty to give Dennis all the help she could. Alice sent a copy of the letter on
to Wayne, and that was the end of it.

But when Mary shared her beliefs with the Privens, they got on the horn to
demand answers. The Peace Corps general counsel's office received a collect call
from Sidney Priven, who said that he was thinking of calling his congressman
about a possible conspiracy by the Tongans to imprison his son. Dennis's brother
Jay followed up to say that the family wondered whether the charges against Dennis weren't a "cover-up" for a Tongan having done the murder.

The Peace Corps also got a call from Dennis's uncle, a retired New Jersey
state judge. The Honorable Raymond J. Stewart, now living in Florida, questioned whether the program was doing enough for his nephew.

These calls made Peace Corps nervous. There was no need for Sidney to go
to his congressman, the chief counsel explained patiently. The congressman would
only call the Peace Corps and ask what Peace Corps was doing for Dennis and
find out that everything that could be done was already being done.

Then the chief counsel undertook to do something about Mary. The chief
counsel was a Nixon appointee called Russell C. Lynch. His background was in
corporate life, as general counsel at a couple of big conglomerates, and he had
learned early on that when scandal struck the first thing to do was to "keep your
goddamned mouth shut and protect the family jewels." Mary wasn't keeping her
mouth shut, Mary was in an exalted state. Following her initial vision in the cathedral, she had had a second visitation. She woke up in the middle of the night with
the sense that evil was present in her room and was struggling with her, but she
then overcame it.

Lynch sent his deputy out to Tonga with strict instructions. Peace Corps was to remain neutral. Officials must avoid "all staff efforts that tend toward helping to defend Priven." Mary was allowed to address Dennis's comfort and conditions and find a lawyer for him, no more.

Associate counsel Paul Magid was the man for the job, a small, laconic man with blue eyes and a dry manner, logical, low-key, and brimming with Peace Corps values. He'd been a volunteer himself. But even Malawi was not so far away as this place. It was going to take him five flights and thirty-six hours to get there.

The world is a seesaw, it constantly needs to be balanced. When someone gets off one end, someone else had better get on. So just as the small dry associate counsel departed Washington for Nuku'alofa, a tall comical volunteer from Nuku'alofa stepped into the American capital.

Never having been east of Denver in his life, Emile Hons walked around Washington with his mouth open, trying to see if the Washington Monument was really there, then making sure it was still there when he got to the next intersection.

Officialdom was not his element, now here he was in a big building across from the White House, wearing his striped missionary pants and leather shoes that he'd shined with butter, riding from floor to floor with a bunch of guys in white shirts and ties, and thankful that he'd gotten another volunteer to give him a haircut before he left.

It wasn't entirely clear to Emile what he was doing here, whether anyone wanted to hear what he had to say. When he started to tell Special Services about the case—Dennis had been feuding with Mary, Dennis had become depressed, now some people were blaming Mary for what had happened, and anyway, what was the policy on transfers?—the Peace Corps staff got defensive. Mary wasn't responsible for Dennis's depression, they said. And besides, the Tongans were fixing to put Dennis on trial in a couple of months. Maybe it was best if Emile kept a low profile.

When he got to the general counsel's office, Emile offered that he had a lawyer friend back in San Bruno who was connected to the American Polynesian community. Maybe that lawyer would be useful. "Oh shit—don't do that," one lawyer said, and another echoed that. "This has been a terrible thing," he said. "We're taking care of it, it's probably best for you to get on with your life."

So Emile wandered back to the National Hotel and procrastinated by washing a shirt and socks in the sink. The lawyers had given him a phone number, NIghtingale-6–3677, and told him he should be careful what he said.

Jay and Sidney both got on the line.

"I wanted to tell you I just saw Dennis a few days ago and Dennis is OK," Emile said. "He said to give you his regards."

Then Jay asked Emile if he really thought that Dennis had killed Deb.

"I wish I had some different answer or some other opinion, but yes, I really think he did."

The men were surprisingly composed, almost as if they knew it already. They thanked Emile for calling, and he told them how sorry he was about the whole thing and hung up, then stretched out on his back on the bed.

For a while he watched the campaign coverage, then he dozed off and a show came on that made him sit up. Women inmates had been mysteriously vanishing from a prison, and three beautiful undercover investigators were sent out to find out what was going on. A blonde, a brunette, and a redhead, they took instructions from an older guy called Charlie who worked in an office. They got themselves arrested so that they could look around inside, and the guards had put them at hard labor, digging potatoes. Their *huhus* strained against the buttons of their khaki prison tops. Emile could even see the shapes of their nipples.

Then they came out of the field into the prison itself and a tough female guard forced them to disrobe and shower. All this as they were quietly cracking the case.

It was pure sex, right on television, and boggled Emile's mind. What had happened to his country while he was going to church and dancing for His Majesty and observing all the *tapus*? He felt himself tumbling down through some milky liberation dream. You could show or say anything you wanted here—except about Dennis, that is.

19

"A Flick of the Tip"

The preliminary hearing began on Monday morning in the police court a half block from the jail. The court's gray benches were filled, and in the afternoon, high school students crowded the back of the room. Dennis sat in the dock saying nothing at all, and rococo stories about him passed through the gallery. He had Mafia connections back in the United States. He had had a wife and children there, and had divorced the wife. Or maybe killed her.

Mary George sat with Tanya Dobbs, while the two men from Washington, the NANEAP director and associate counsel, sat toward the back of the courtroom. Jack Andrews took copious notes, and Paul Magid raised an eyebrow. Jack was not a lawyer, neither was Mary. They were program officers, charged with running normal Peace Corps operations. Headquarters had instructed them through Magid to stay out of the criminal case. "Neither you nor other Peace Corps staff should become involved in supplanting, supplementing, or otherwise attempting to assist Priven in those areas where specialized and professional help has been supplied to him."

That specialized help was now in an anteroom.

When the door opened on Clive Edwards, a person's first thought was that he was an orderly, a driver, an underling, someone permitted his daytime sleepiness. His big shoulders were tilted, there was a hint of fond stupor in his eyes. He mumbled when he spoke. Then the words began to string together, and the listener abruptly became aware that the man was cunning, for he was able to convey complex thoughts in simple words.

Tongans called the lawyer Neti, or Ned (from Edwards), and they were proud of him, and fearful. After the King, he was the most famous Tongan in New Zealand. He was a kind of raceman. He fought for Tongans, and fought dirty when he needed to. He was actually descended from one of the mutineers on the *Bounty*, and had graduated in the legendary first class of Tonga High School along with the police minister 'Akau'ola. But when 'Akau'ola was summoned back from New Zealand by the King, Clive had stayed on. He defended Tongan boys who ran afoul of the law, and had been elected to the Auckland City Council, where he spoke out against the government's hideous dawn raids on Tongans who had overstayed their visas.

From the start, he and Dennis did not get along. Dennis complained that Clive didn't have time for him, and as for Clive, he said that Dennis was a sphinx.

Twice he had refused to talk to him. The first time, Clive stopped into cell 1 to introduce himself.

"I am Tongan but I have a British passport, I have done many criminal cases in Auckland. These are serious charges, and I am here to defend you against them."

The man said nothing back, not a word, hunched in the cell, gazing away. It occurred to Clive that Dennis was racially prejudiced, and he reported as much to the Peace Corps.

"I don't think he wants a Tongan as an attorney. That is all right. If a client doesn't want me, I don't wish to represent him."

But Jack and Mary insisted that Clive stay on. Jack was shrewd, Jack pronounced him first-rate, and Mary knew that it was good to have a Tongan hero standing before the jury.

On his second visit, Clive brought a legal pad.

"Dennis, I must prepare my case. I will have to counter what the police say, so I need to know everything about what happened that night in your words. Then I will be able to pick apart their story at the preliminary hearing."

Again, nothing. Clive repeated himself and flapped the pad against his leg. The young man stared off into space.

◆   ◆   ◆

Now his tilting bulk was invested in a black robe, and Clive drew a horsehair wig from a leather-covered box and went into the court and proceeded to perform for his new employer.

"A woman died that night—" Faka'ilo testified.

Clive rose to his feet.

"I refute the statement made by this witness. He did not see what happened. He only heard about it. Your Worship, you are well aware, in law no one is allowed, it does not matter whether he is a policeman or not, to say that 'So and so is dead' when he only heard about it."

"Counsel is right," said the magistrate, Hingano Helu, a bookish man with glasses. "Keep to the police side of it."

Faka'ilo began again more cautiously.

"I went with other officers to hospital. Dr. Puloka was there. He showed us the body of Debbie Gardner. I stood beside her and found that she wasn't breathing. I then went to the scene of the crime. I examined the house. I found a syringe there. I also found a bottle outside."

And Christine Oldroyd, the English prosecutor, produced exhibit C, the jar with the bottle inside.

"Yes that is the bottle. It has some liquid in it—"

Exhibit D. One flip-flop only.

Exhibit E. A dagger.

Exhibit G. A pair of spectacles.

Exhibit H. Torch.

Exhibit I. A piece of pipe.

But Clive could take only so much of this before his chair grated harshly on the floor and he rose again.

"I am tired of these answers. The inspector has done nothing to connect any exhibit to the accused. Your Worship, this is no joke. What happened that night is a life was lost. And now—a life is under judgment."

After the first day of the hearing, the Bird Lady and Laura brought Dennis's food to the jail, but an officer was haranguing the constables and the women apologized for interrupting and set the food down before going back outside to wait to see Dennis. The harangue went on for a while, and Laura left, the ice cream melted, and the Bird Lady had to go to John's Takeaway in the King's Road to get more. When she got back, Dennis waved the Bird Lady to a seat beside him.

He was quite wound up and dragged out the meal for nearly an hour as he talked about all the holes in the Crown's case. The Bird Lady was happy to see him in such a confident mood, such a departure from the daunted way he had discussed the evidence a week before. There was nothing to establish he committed the crime. No one saw him do it. There was no motive established. And no confession. None of the objects found at Deb's hut were shown to be his. No one was sure if there was anyone else there, and as for those eyewitnesses, they had been far away and in darkness.

"In the States, this case would have been thrown out long ago," Dennis said.

The associate counsel also visited Dennis in jail, and found him to be perfectly lucid and rational, though full of complaint about Mary George. She was unyielding and incompetent. Magid needed Dennis's signature for a form authorizing the Peace Corps to release information about the case to his uncle the judge. Dennis agreed, but he told the lawyer that he did not want his parents coming out to Tonga. Though if he was convicted, he would get his friends to write their congressmen.

Magid asked after his health and Dennis showed him his wrists then lifted his shirt to show him a superficial wound on his chest, now scabbed over.

"From a flick of the tip," Dennis said.

A flick of the tip—that wasn't the sort of thing that generally happened when you were cutting your wrists. Evidently, someone else had caused that, sent the tip of the Seahorse into Dennis's own chest. Magid made note of that on his legal pad.

So again, it was that parallel Peace Corps investigation. Peace Corps was gathering its own facts, *pālangi* facts, facts that would never be turned over to the Tongan authorities. The associate counsel had also picked up Mary's concern about drugs coming up in the trial, and he visited Deb's house and made a casual search. He noted a bloodstain on a paperback, but he couldn't find any drugs.

◆   ◆   ◆

On the second day of the hearing, the prosecution scrambled. Clive had found the biggest hole in the Crown's case, the failure to connect the physical evidence to Dennis, and Christine called a student from Tupou High, who made even more of a mess of things.

"Do you see Dennis Priven?"

"Yes, that's him standing there, in the dock."

"Is this his torch?"

"I've often seen him with a torch. I cannot confirm that this is the one."

"Glasses?"

"Yes, he often wears glasses but I cannot confirm that this is the one."

"A knife?"

"Yes, he gave me his knife last year to use in experiments, but I don't remember what make or type it was. I don't know anything about this knife—"

Dennis laughed, or coughed, and Christine glanced over at him and was not sure whether he was mad or if it was nerves. She called another few witnesses before Clive moved that this mishmash of insinuation be tossed out of court.

"Your Worship, the defense is very much troubled over the way in which the prosecution has presented its case. No one claimed to have seen the accused inflicting the wounds. Therefore the prosecution had to prove that there was no one else in the house on the night in question. That is the only foundation for the prosecution's case, and it failed to do so.

"Exhibits have been brought before the court, but there has not been any lawful evidence whatsoever as to whom they belong. All the inspector says when I cross-examined him about the exhibits was that he would prove it later."

Helu was unswayed. He said that Clive was right about the police investigation, it should have been much more thorough, but the eyewitnesses who'd claimed to see Dennis in Ngele'ia had made a compelling case.

"I leave the accused in the hands of the judge of the Supreme Court—" he ruled, and left the court.

No one was really surprised. Christine was formal, maybe even prim, bespectacled, with strawberry blonde hair, but she understood the moral drama of the case. She possessed cold fire, and brought out her most important exhibit late in the proceedings, when Lutui took the stand and said that he had treated a woman named Debbie Gardner that night.

"And when did you see her?"

"I first saw her at about ten minutes to ten."

"What was she wearing?"

"A long dress."

"I want to show you exhibit N—"

Now Christine unfolded a plastic sheet, and what it revealed silenced the court, a crushed shroud with feminine darts and gatherings, its long swirling folds flattened and stiffened with purple black blood. Only small patches of the dress's shoulders and skirt were still white, and Dennis jerked in the box as it was spread out and strangled a noise in his throat, and rolled his big head around his shoulders trying not to look.

It was not only the sight of the dress. The blood stank. People in the courtroom told themselves it was the market next door, or a wind bringing sewer gas. No, it was darker and drier than that, gagging, hellish.

Christine and Lutui calmly pointed out all the jagged knife holes in the dress, in the bodice, sleeves, skirt, then the dress was rewrapped in the plastic, still the smell lingered, and some Tongans in the court made quiet prayers invoking the blood of Jesus.

After the second day, the police did not let the *pālangis* sit with Dennis when they brought him his dinner. The officers were resentful. Clive had made the brass look like fools, and their Minister was enraged by all the gaps in the case.

"You really have to go," a policeman said, shooing the *pālangis* outside. "You cannot talk to Dennis, you just have to go!"

Three of them duly went outside, but Laura was wily and managed to stay inside, and though the police looked to Paul to order his wife to leave, he did not. So she gave Dennis his dinner, and this time the ice cream did not melt.

The friends regrouped later at the Bouchers' house. Clive was very good. He had made mincemeat of Christine and discredited nearly every witness. There had been no end of police foul-ups and confusion. Yes, Dennis's clothes were bloodstained, but the prosecution had not analyzed the stains and shown them to be blood, let alone whose. That blood, if it was blood, could have been Dennis's. After all, he had cut his wrists. The *tapu* jar had been introduced, and the police had declared the liquid to be poison, but they could say nothing about it definitively. And anyway, what did poison have to do with a stabbing?

As for Dennis's former student who could not identify the knife or glasses, a person could not help but be amused. Some *pālangi* teachers had seen Siua later at Tupou High and talked to him about his testimony at the hearing. He'd shrugged, said he didn't know anything about the case.

The Bird Lady and other teachers had smiled approvingly at Siua and told him, "That is the best thing to know in a case like this."

Notwithstanding the magistrate's ruling, Dennis was heartened. Laura reported that as he ate his dinner, he said repeatedly, "Seven weeks and two days." The trial was to begin in the Supreme Court on December 13, and it was not likely to go on for more than a few days before this business was brought to an end and Dennis could be sent home.

Oh I hope so, the Bird Lady said. It was probably just as well to wait seven weeks, to let emotions die down. The Crown was sure to coach the witnesses and get them even more confused. "It'll be hard on Dennis, though," she wrote her parents.

When Clive spoke to Laura about the timing, he was more restrained. "Your friend will be gotten out of here. But it will take a long time."

Clive was leaving for New Zealand, but first he paid a third visit to his puzzling client.

"Now, Dennis, I must review the allegations that have been made against you and get your response."

The lawyer looked at a pad. That Dennis had bicycled to Deb's house on the night of October 14 and stabbed her twenty-two times and left a flip-flop and his glasses and syringe, et cetera.

"The evidence while circumstantial is actually quite strong. Now I must go back to Auckland and build your case. I need as much information as you can provide in order to do so."

Again, nothing. Clive waited an interval as Dennis stared off into space, then stepped from the cell with his legal pad empty. Well, the man is insane, he thought, that is obvious.

Clive's withering cross-examination of the witnesses had actually been something of a charade. He wanted the Crown to believe that he was preparing a defense of not guilty on the facts so they would waste time trying to bolster the evidence that Dennis had been at Deb's hut that night. "I have sidetracked them," he bragged in a letter to the Peace Corps general counsel.

An insanity defense was really the only option. Washington was on board with the plan. The associate counsel felt the case against Dennis to be open and shut. He had heard Dennis's statement about "a flick of the tip," while the regional director cabled the Peace Corps director to press him to get payment to the psychiatrist in Hawaii.

SUGGEST URGENT ATTENTION TO THIS AS ABSOLUTELY ESSENTIAL WITNESS.

To be sure that no one in Tonga got wise to the defense plan, Jack had his cables hand-carried to Fiji and New Zealand and sent from there to headquarters.

The associate counsel was leaving, and vowed never to return. He'd expected a tropical paradise, he'd found a flat island where pretty transvestites floated up outside the leading hotel offering to perform fellatio on him then and there. Deb's house was grim. They wouldn't have put a dog in a house like that back in Malawi.

Mary and Mike Basile gave him a ride to the airport, and being a logical sort, Magid took the opportunity to remind Mary about Peace Corps policy. It was OK to make sure that Dennis was treated well in jail, that he was in good health and had what he needed, and to convey messages to his family. "But beyond that Peace Corps really has to be neutral. There's a fine line between making sure that Dennis is defended and actually participating in that defense."

Mary was way past that sort of hairsplitting.

"I'm trying to see that Dennis's human rights are observed."

"The police feel that you are being disrespectful."

"The police! The police are incompetent and cruel."

"Well, that may be true, but they have a job to do and you could actually be hurting Dennis by annoying them. The police have seen some of your actions as interfering."

"The police are thugs, they are stupid oafs and buffoons—"

The two former volunteers in the car were silent, and Mary went on for some time—calling the police idiots, goons, and many other epithets of that sort. The lawyer opened his mouth only a couple of times, to say he was not giving Mary his impressions but conveying the impressions of the police, but before long he shut down entirely. He had come to see Mary as a hysteric. And though Magid was not the only one at headquarters who was critical of Mary—she had been known as "Miss Pink" during training because of her coordinated outfits—she seemed to have powerful protection.

The NANEAP director had come out to Tonga to help a middle manager through a nightmare. Jack had turned 57 in Nuku'alofa that week and though he had never dealt with something quite like this before, he was resourceful and attentive.

So attentive that reports went out about Jack and Mary on the kerosene radio.

It was said that he was seen in Mary's pretty gardens on the foreshore, walking with her by moonlight, holding her hand. Well, Mary was a very touchy person. A manager used different techniques with different underlings, and Mary seemed to require that type of comforting. Maybe she had reached for Jack's hand—

Whatever the reason, Jack gave Mary a gold star when he reported to his boss, John Dellenback.

STAFF EXHAUSTED BY TRAGEDY AND LONG HOURS BUT BEGINNING
TO FEEL BETTER. MARY HAS HANDLED SITUATION SUPERBLY.

The police minister grew angered if a subordinate disturbed him on a trivial pretext during meals, but if the case warranted he might be approached. So it happened that on the second day of the preliminary hearing, an officer came to the patio and dropped a newspaper before 'Akau'ola, and now *Rex v Priven* was an issue of national sovereignty.

The October 26 *Fiji Times* was the first newspaper outside the Kingdom to report that Deb Gardner had been murdered. And it had misreported the story. "A young Tongan has been charged in Nuku'alofa with murdering an American schoolteacher," the article began, and went on to cite unnamed Peace Corps sources for its news.

Every Polynesian knows the history of *pālangi* mischief in the far reaches of the Pacific. The battleships from different nations showing up in harbors in the Marquesas and Samoa to fight it out for empire; the exchange of nails and gasoline for women; the use of islands as nuclear targets and waste dumps; the Navy man's wife who delayed Hawaiian statehood by falsely claiming that five Hawaiians had raped her, and her husband and mother who had killed one of those Hawaiians and got off thanks to Clarence Darrow. . . .

Now mischief was afoot in Tonga.

"A young Tongan has been charged . . ."

The Crown prosecutor, Christine Oldroyd, had pushed 'Akau'ola to have scientific tests performed on the evidence gathered from Deb's hut. At first the minister had been embarrassed by the request. Oldroyd was spoiled by England. Little Tonga lacked the facilities to conduct such tests, and John McMath had turned the police down. But now the Americans had called the minister's bluff, and 'Akau'ola, who was given to nationalistic bluster, who liked to go on about the Tongan character and say that a Tongan warrior would put his life on the line as casually as an American would sip a cup of coffee, 'Akau'ola turned to the most powerful ally he had.

He called his old sergeant in the Auckland police and asked whether the Kiwis might do some tests.

Could the blood that had soaked the girl's dress be linked to the blood on the boy's shorts? And what was in the jar that Dennis said was *tapu*? Was there blood on the syringe?

The New Zealand government agreed. The Department of Scientific and Industrial Research had a chemistry division, it was willing to perform the analyses for free. So a couple of days after the preliminary hearing, the police sent off a box of exhibits to New Zealand.

GRATEFUL EXPEDITE ANALYSES MURDER EXHIBITS. THERE MAY BE DRUGS. ACCUSED WAS A CHEMISTRY TEACHER.

Dennis's jeans cut-offs and his long-faded orange Illinois T-shirt went in the bottom, and Deb's stiffened dress was folded in as a kind of pillow for the various items that followed. Metal pipe, syringe, Seahorse knife, a bottle containing crystals, and of course the bottle inside the jar with the unknown liquid inside it.

If the box was lost, so was the case. A police officer carried it to the pilot of the Thursday morning Air Pacific flight with clear instructions that the pilot surrender it only to Constable Mitchell.

The Tongan police had done a lousy job packing it. En route the jar somehow bumped or turned over, so that the bottle spilled out its contents over other pieces of evidence, smearing the written labels. The result was that when a chemist opened the box in a laboratory at the Department of Scientific and Industrial Research in downtown Auckland, he did not really need to perform any analysis at all on the bottle, for the unpleasant smell popping out of the box, in a manner of speaking, Dennis's genie, was as strong as bitter almond and unmistakable.

Nelson scribbled, Cyanide.

20

Once Out, All Out

A week before the presidential election, Kelly Downum drove much too fast down through the mountains from Boulder to Denver.

After washing out of Peace Corps ten months before, Kelly had moved back to Colorado and started driving a truck for the *Rocky Mountain News*, and when Vic Casale washed out a few weeks later, he called Kelly and Kelly said the paper was hiring. So Vic followed him out there, and for a little while they were both living in Denver and making deliveries for the paper, till Kelly moved on to graduate school in Boulder. The truth was Kelly couldn't keep up with Vic in the living arts. Kelly was studious and a bit of a stick-in-the-mud—his future was in plant science—while Vic was a hard and glamorous liver, could stay up all night and go to work the next day.

But on this day any distance between the friends collapsed.

"Vic, I need to talk to you."

"Sure, come to my place."

And as Kelly drove south in the mountains, the Kingdom unspooled in his mind. What was that about? Some sort of escape or interlude, the French foreign legion, Kelly did not know, except it was crazy and lost and now he couldn't hold on to it even if he wanted. Lampshade Phil bringing him into a bordello by the sea and announcing to some beautiful transvestite at the door, "This is Kelly, he's a friend, give him one without the disease—" What was that? Well, now it was twisting away forever and taking something of him with it.

The two friends crossed Lincoln Boulevard to the Japanese Kitchen, and after the woman poured two Kirins, Kelly gathered himself and Vic looked into his friend's face and saw only devastation.

"Vic, Deb is gone."

Vic's green-brown eyes flicked away. In an odd way it was as if he already knew. He took a breath.

"Man—Kelly—what do you know?"

"Here, this is from Cade."

Kelly got the letter from his shirt pocket, smoothed it out on the bar. Cade Campbell had shipped out with them, the one African American in Tonga 16.

"*Mālō e lelei* Kelly," Cade began, her script long and sweeping, and tried a few sentences of small talk before giving up.

"Think of you often. I have a feeling you have not received all my letters . . .

"Well, one thing that spurred me on to write today was, I am very sorry to say, some very upsetting and sad news. On Thursday night past, Debbie died of stab wounds received at her *'api* during a violent quarrel with Dennis who, I think, came with Tonga 14. Although all is still piecemeal, I shall try to give you all I know—"

Vic's breath choked in his throat. It was like someone had kicked him in the chest, still he didn't cry, he wouldn't let that happen, but he covered his eyes and thought of the bad way he and Deb had left off.

Cade offered a scene from Ngele'ia that night: A Tongan boy had seen Dennis run out of Deb's *fale* and ride away on his bike. And she described the dance at the Dateline: Debbie had fallen on the dance floor and just stretched out there while her partner gathered her up. She was oblivious to the people staring. Cade wasn't sure whether that had any bearing.

The men got another round and drank a toast to Deborah Gardner, and Vic called Wichita, who was on his way to becoming Dr. David Scharnhorst, and the idea that Deb had been killed, Deb who was nothing but a good kid, a kid from a small city and a state school, someone who just wanted to help, it tore away the last good feeling Wichita had about Peace Corps, he was never going to reapply to John Kennedy's program.

He thought about the idealistic questions regarding other cultures he'd answered on the long government application. He thought about him and Vic and Deb standing out back of Sela's memorizing the first two pages of Richard Brautigan's "Watermelon Pickle," line by line. He thought of Deb leaning over to him in her soggy sundress as he moaned about his troubles. Why? Why had anything happened there?

So it was in this way that two grad students in science and a newspaper truck driver learned about a two-week-old international story that no one in an American newsroom knew anything about.

From the time that it had learned of Deb's death, the Peace Corps had been determined to control communication about the case, and to a surprising degree, control is what it had gotten. The only way that anyone in the American public knew about the murder was from volunteer letters. "I thought I had to tell you guys about this, because you may have read it in the papers and started to worry," Greg Brombach wrote to his parents four days afterward. They hadn't.

The Peace Corps manual stated that the Office of Public Affairs was to prepare a press release about a volunteer death immediately upon learning of it. That policy had been ignored, and the few people who knew anything about the case at headquarters had practiced a type of deception in their internal communications, carefully avoiding the word murder. A day after he'd learned of the murder, Peace Corps director John Dellenback had been evasive about it in his weekly memo to Action director Michael Balzano. A week later he continued to be evasive:

"Twenty-three-year-old PCV Deborah Gardner died in Tonga on October 14, the 11th Volunteer fatality this year."

Vice President Nelson Rockefeller got the same line. The Office of Special Services always sent the vice president a memo after a Peace Corps death so that he could express his condolences to the parents, and these memos always informed the vice president of the cause: "Hemmorhages caused by a motorcycle accident," "a heart attack," "an overdose of barbiturates."

The memo about Deb didn't go into any of that: "She died shortly after her arrival at the hospital."

Ten years before, Warren Wiggins had hastened to inform President Lyndon Johnson and Vice President Hubert Humphrey that a volunteer in Tanzania had been charged with murdering a fellow volunteer. The White House took an interest in the case, and soon Congress did too.

It was not hard to imagine the White House taking an interest now. If Jerry Ford had anything, he had the common touch. Several times he'd become involved in international cases, once asking the Mexican government to furlough an American prisoner so that he could visit his dying mother. Besides, Gardner was a family name. The president's son was named Gardner, his mother was Dorothy Ayer Gardner, born in Illinois, as Deborah Ann Gardner had been.

But Ford was not informed. He was in a race for his political life.

It is surely doubtful that the murder could have become an issue in the race, but no one in politics likes surprises, and John Dellenback was then out on the campaign trail in his bow tie, saying that "Ford has a proven track record" and voters couldn't trust Jimmy Carter, and his volunteers were striving for world peace by building ties between people. He didn't need Deb's picture in the paper alongside reports that a fellow volunteer had brought cyanide to her house then stabbed her twenty-two times.

Dellenback had recently assured Congress about volunteer quality. The bad ones were weeded out at stagings before they could even leave the States, he testified. There was "a careful discussion with this person—sometimes a couple of hours—about whys and wherefores." In the field, volunteers were regularly checked out by others so that emergencies did not arise. If there was trouble, they could be medevac'd.

Selection was a soft spot. The House Committee on International Relations had lately raised the issue in a highly critical report. "Preliminary screening procedures within the United States are now practically nonexistent," it said, leaving the task of weeding out unacceptable volunteers to the overseas staff, which was already overburdened. Peace Corps had gotten the same complaint from the head psychiatrist at Sibley Memorial Hospital who fielded psychological evacuations. He told Peace Corps there were far too many psych-evacs, Washington ought to be doing more to screen these people.

How had Dennis Priven gotten in and hung around?

He'd been seen briefly in the States by a psychologist who'd asked the male volunteers whether they were going to be able to handle two years of celibacy. Then was sworn in by a country director who was also a psychologist, who subsequently medevac'd him back to the United States for strange stomach pains doctors had not been able to figure out. He had set a big knife on the sill of the blackboard before the start of chemistry class, cut his door in half and painted a face on the door, showed up morning after morning at Deb's high school even after a vice principal had told him not to, caused the Peace Corps secretary to fear him, and had a screaming match with the country director days before the murder. The murdered volunteer's mother and the Peace Corps secretary, Makaleta, said that Deb had expressed fear of Dennis, while the chargé d'affaires had reported to the Ambassador that Dennis had been taking a "tranquilizer" for several weeks before the murder, and Mary said in a cable to Washington that she had discussed the "circumstances leading to the allegations" in a telephone call with the country desk officer.

What did Peace Corps know, and when did they know it?

Congress was supposed to be informed about important cases. That was also in the Peace Corps manual. The two staffers to the House Committee on International Relations who dealt with Peace Corps never heard word one about Deborah Gardner.

In one of the bloody coincidences of journalism, as Peter O'Loughlin would later have it, a veteran newsman had happened to be in Tonga the day after the murder.

On Friday, October 15, O'Loughlin, an Australian who had just taken over the Associated Press bureau in Sydney, was on an introductory tour of the Pacific to seek out dependable correspondents, when Paua Manu'atu, the editor of the *Chronicle*, told him of the murder the night before. O'Loughlin had spent years in southeast Asia, he knew a good story, and he got a cab to the lagoon, where he encountered Mary, who made it her business to avoid him. The police helped him, though, and he telexed a story about the case to Sydney from the International Dateline Hotel.

Then by some other bloody coincidence of journalism, that story was never published. Apparently there was a mix-up between the international and domestic wires at Rockefeller Center in New York. The story was chiefly of interest to a domestic American audience, but it came in to the international desk, which failed to pass it along.

Nonetheless, the Associated Press now possessed some knowledge of the case, and twelve days later, after the *Fiji Times* published the first report on the murder outside Tonga, an AP reporter in Washington called Peace Corps headquarters to ask what it knew.

Peace Corps said it had a man in the South Pacific, looking into the case, it would get back to AP when there was something solid.

So the story was coming out, it was just a question of when it would break in earnest. A small delegation knocked on Mike Balzano's door. Russ Lynch cleared his throat, spoke in his beautiful deep voice.

"Mike, we have a terrible case, a volunteer in Tonga has been arrested for killing another volunteer, a young woman. He used a knife, they have the death penalty there, and now Peace Corps is arranging his defense."

"What? Why?" Balzano said.

Lynch was taken aback. "Why? Well, Mike, they want to decapitate this guy."

"So what?"

"Well, he's a volunteer. The question is how to get him off."

"No one's getting him off, let them behead him. You say he did it, what's so bad about that? That solves the problem."

The lawyers looked at one another, and Lynch took a different tack.

"Mike, it's the law. When a volunteer gets into trouble, we have no choice. Congress says we have to pay for his defense."

"I see, so we're just giving him due process," Balzano said, and the suits nodded and slunk back out.

Balzano didn't follow up on the case. The mistrust between him and Peace Corps was too profound. Still, it is interesting to consider that fifteen years after Peace Corps was founded on the principle that volunteers should live as their hosts lived and suffer the consequences of their actions in host country courts with host country judges, the only person at 806 Connecticut Avenue who was sticking to it was thought to be Peace Corps's greatest enemy.

The associate counsel got back to Washington that weekend after another five flights and thirty-six hours and came into the office Monday with a bunch of papers from the Kingdom. Magid called Mrs. Priven to tell her Dennis was OK, then sent Judge Stewart, Dennis's uncle, a thirty-page transcript of the preliminary hearing in Nuku'alofa, detailing the testimony of police and eyewitnesses. He briefed Public Affairs, too. And the next day public affairs called the AP back with a paragraph of news.

Tuesday, November 2, Election Day. Fully nineteen days had elapsed since Peace Corps learned of the events described in the release.

At the building across from the White House, Wednesday, November 3, was an electrifying day. For eight years, Peace Corps had endured live burial, now its long Republican nightmare was ending. A Democrat was coming into the White House and not just any Democrat, Jimmy Carter's mother had served in Peace Corps/India. One staffer unfurled a bedsheet from a window overlooking the White House with a taunt painted on it, WE WON.

Election news dominated the papers that day, and though the leading newspapers could find no room for the Tonga murder, it did appear in many other papers throughout the country. The *Washington Star* ran it, and the *New York Daily News*. So did the *St. Louis Globe-Democrat*, the *Charlotte Observer*, and the *San Diego Union*. And of course, newspapers in Seattle and Tacoma also published it.

"A Peace Corps volunteer has been charged with murder in the death of another volunteer on the South Pacific Island of Tonga, Peace Corps said today. Dennis Pervin, 24, [sic] of Brooklyn, New York, was charged by Tongan authorities in the Oct. 14 slaying of Deborah Gardner, 23 . . ."

She'd been slain was all, not a word about how. Reporters from the *Tacoma News Tribune* and the *Tacoma Weekly Review* called Peace Corps and were passed along to Magid.

"How was Ms. Gardner killed?" the weekly's man asked.

"She died of multiple stab wounds to the upper body."

"Had there been a fight?"

"I have no knowledge of that."

"Were the two volunteers living together?"

"Negative."

"What evidence was presented at the preliminary hearing?"

Magid was circumspect. He said that there was medical evidence about the cause of death and evidence of items that were found at the girl's house. But he did not detail the items, syringe, needles, and metal pipe, nor tell the reporter about the eyewitnesses. The transcript of the public hearing had been sent to Dennis's uncle, the judge, but was not given to reporters.

Peace Corps had a very limited role, Magid explained. It had appointed an attorney to defend Dennis but beyond that it would be strictly neutral, monitoring the case to make sure that Dennis got a fair trial.

A couple of reporters in New York called Peace Corps to get the Privens' number, but Russ Lynch said the Privens weren't interested. The Associated Press went further. There were half a dozen Privens in the Brooklyn book, and an AP reporter called the co-op in Sheepshead Bay and said, "I'm hoping to reach the family of Dennis Priven—"

"You have the wrong number," Mimi said, and hung up the phone.

Reporters also called Peace Corps to get the Gardners' phone number. A man at the *Seattle Times* wanted to do a feature story about the family. When he learned about these overtures, Russ Lynch went to Bette Burke in Special Services on the eighth floor to give her some advice to pass along to Alice Gardner, and after lunch, Bette made careful note of the general counsel's advice.

"Russ said, Other parents avoiding discussing. Allude to fact, *Once out*, all out."

Did the Gardners have something they didn't want to come out?

Of course people were saying similar things in Tonga. Half the guys in the country had been hitting on Debbie Gardner so that made her the village bike, Debbie Gardner's house had been searched for drugs and other suspicious material, Debbie Gardner hadn't had the modesty to wear underpants to her own murder.

Russ Lynch's threat wasn't necessary, Alice had no desire to talk to a reporter. But in issuing it, Lynch had a powerful ally on the other side of the world.

Judge Hill was the new piusne judge. He sat in Tonga's Supreme Court. *Rex v Priven* was to be his first major case.

A red-faced round-headed Englishman who had grown up in the East Africa colonies and became a major in the army, before boldly starting a second career back in London as a barrister and later a recorder, a type of judge, Henry Hubert Hill was 57 when he saw the Crown Agent's advertisement for a judge in Tonga and felt called back to the tropics.

Dennis's lawyer, Clive Edwards, had made a point of meeting Hill, sussing him out, and Hill had offered a comment on the case.

"The alleged victim was of loose moral character. It is too bad that the defendant was involved with her. And she was a drug user."

As things were to turn out, Henry Hill was on his way to going troppo. "Going troppo" is the opposite of "island fever," it means getting too well adjusted, losing one's civilization. Wearing a grimy sarong, say, or having a second family with a servant girl. Going troppo was scandalous when it befell a respectable man, and Hill was the highest legal official in the Kingdom. In months and years to come, Henry Hill drank too much and sent his houseboy Tausili to bring home women who caught his eye, the first of whom was disputatious and grabby. But the second one was mild-mannered and languorous like a girl in a Gauguin painting, and she threw the first girl's clothes out of the closet and moved in and later bore a half-*pālangi* child she told people was the judge's. The judge denied that, but she named the girl after the judge's beloved cousin, and by then he was a full-blown scandal. *Fokisis* came to the courthouse saying he owed them money. One time a *fokisi* came to the door when the court was in session, and the judge gaveled the proceedings to a stop. After that someone distributed an anonymous cartoon to members of the Cabinet depicting Hill as a bat in robes being chased through the air by flying *fokisis*.

Privy Council did not extend his contract, and Hill went back to England, and wound up alone, in Hammersmith, with a bad dose of Alzheimer's.

But that is way ahead of this story. At this point, only one girl had come to the judge's house, the grabby one. None of the rest had happened.

Meantime, though, Henry Hill had passed judgment on Deb Gardner, and Clive passed the word along to Magid, who carried it back to headquarters in his legal pad, where it struck the chief counsel as good news. Or if not good, certainly useful.

21

Wayne's Appeal

At the end of July 2001, I crossed a clanking bridge over the Clearwater River in Idaho and went up a winding road past a tangle of blackberry bushes laden with giant berries, and a new red Ford 250 pickup came down. The truck pulled over to his side, I pulled over on mine and ran across the road.

Wayne Gardner had a house a few miles back in the Clearwater breaks. He lived on the ground floor by himself. The family upstairs, the Gregors, had joined him from Alaska. They brought him his mail and were going to keep an eye on him till he was room temperature, he said. It was like a hunter's cabin but with all the amenities. There were trophies everywhere, and a gun room full of guns. Wolverine skins, sheep heads. Lynxes hanging from a coatrack, turkey beards, bears. Here and there on the wall were mounted hunters' gags, and a doormat said, Protected by Smith & Wesson, which was no gag at all.

That afternoon we sat outside by a golf green Wayne had built, and I told him what I was doing. The apricot trees were filled with apricots, and from time to time I took one and ate it and spun the stone into the air down the hillside. Green hummingbirds buzzed by us, and Wayne listened quietly for a while.

"I'll help you any way I can. Now how will you help me get justice for my daughter?"

I told Wayne justice was fine, but I wanted to tell the story. I was a writer, others could deal with the question of justice. On the telephone before I'd come out, Wayne had asked me to determine what the legal consequences might be to Dennis of his actions so many years before, and I'd spoken with a former federal prosecutor who said there was little likelihood of anything happening. The red light had come on in too many ways.

Wayne told me how he had learned about my project. He'd turned 70 in May. A month later he'd driven to Alaska to visit his son, and Craig said, "Dad, there's your mail on the shelf." So Craig had done as I'd asked him fifteen months earlier. But Craig didn't want to reopen the case, and Wayne didn't try and tell Craig what to do. There was no point in trying to twist a Gardner's arm.

Then Wayne had gone to the Tok airport to pick up his second former wife, Barbara, because they were going to drive back to the Lower 48 together, she wanted to drive the Al-Can Highway. "You'll never believe this, there's a writer who wants to do something about Debbie's death and the government," Wayne said, and dropped my letters in her lap. Barbara said, "Wayne, the whole family—everyone's been praying for you for years, that something like this would happen. That you would come to a closure, or that you would be able to open up to all of it."

When Wayne got back to Idaho he called his first former wife, Alice. Twenty-five years had passed since they'd last spoken at their daughter's funeral. Alice didn't recognize Wayne's voice.

"I want to talk to this writer. Is that OK with you?"

Alice said, "Yes."

Wayne wanted to know everything that I had learned in Tonga, he wanted to see all the pictures I'd collected. I showed him everything but the police photo of Deb's dress. By then I'd seen him cry. He would fumble with a cigarette, busy himself with a gun, or pretend to wash a glass at the sink.

I asked him for a letter I could give people saying he was in favor of what I was doing, and Wayne gave me a handwritten letter and said he wanted me to use my computer to put a picture of the Seahorse knife at the top of each copy. I told Wayne I wasn't putting that knife on the letter.

"You're sensitive, I'm not. The truth isn't always pretty," Wayne said. "I worked in sales. You shock them to get their attention, then they're listening and you say what you have to say. You type that up and put in the right punctuation. Or get your wife to do it."

The letter said:

"This is an open letter to all who knew or worked with my daughter, Deborah Ann Gardner. I am her father, Wayne L. Gardner, and I am asking, no, begging you to talk, write or communicate in any way all that you knew about Debbie and her cold blooded murder by Dennis Priven to Philip Weiss. Remember, she was stabbed 22 times by pictured knife . . .

"I know some of you may think that Phil is doing this for the money and fame, I am not denying this may happen, but from my standpoint he is the conduit that I hope will force my government to honor the freedom that was brutally taken from her by a mad dog killer, who today and for the last 25 years has been allowed to follow his dreams by being free. Remember Debbie's dreams and where they are now.

"Please help & talk to Phil. I need you. From a father who still cries a lot."

A lot of people were moved by the letter. One night at Frank Bevacqua's house, Frank read it and shook his head and handed it to Doug Jackson. Jackson's face flushed. "Oh, my." They were surprised by how fresh Wayne's feelings of rage and grief were, but those feelings were also clarifying. Jackson had been wary of my project. After he read the letter he said he'd help me any way he could.

The letter didn't work with others. Two years before Vickie Redpath had brushed me off with a 3-by-5 card written out in block letters:

I'M NOT INTERESTED IN TALKING WITH YOU NOW OR EVER. STOP TRYING TO CONTACT ME AND MEMBERS OF MY FAMILY. V.R.

Wayne's letter didn't alter her stance. Bill MacIntyre, a Tonga 11 who had been willing to meet with me, demurred after he got the letter. He was a religious man, he said I was stirring up a lot of demons in Wayne and spurring his desire for vengeance. Wayne would be better off seeing a pastor or psychiatrist than a journalist.

I told Wayne what Bill had to say, and he took issue with it.

"There's nothing a psychiatrist or a person of the cloth could tell me that I don't already know. Not one of them can tell me why Debbie died. Not one of them can tell me why it was covered up. Not one of them can tell me why some people won't talk to you."

Wayne asked for a copy of all the documents in the case and I made up a big box to go to Idaho. By then I'd done several Freedom of Information Act requests, and the Peace Corps, State Department, and National Archives had released hundreds of pages of documents, including the page with Russ Lynch's threat for Alice Gardner: "*Once out*, all out."

The status of the documents seemed to perplex the government. The case was historic, the only time in forty years and 170,000 volunteers that a Peace Corps volunteer had been found to have killed someone. But officials seemed to sense that it wasn't purely historic, it had an ongoing life. I spent many days in the Archives' glassy tranquil research room in College Park, Maryland, but I knew when my last day had come. For a while I'd had the run of big cardboard boxes containing Peace Corps files. Then I opened a box labeled "Peace Corps Director's Files" and found it completely empty. Every folder inside had been removed, with yellow slips saying the archive had been pulled on grounds of personal privacy.

It was the same with the next box. It too was empty. I did not complain. I felt that I'd stolen a march on history.

The Kingdom wasn't nearly as good as my country at keeping records (the clerks always blamed the 1982 hurricane), but the Supreme Court and police and prime minister gave me just about anything I wanted, and nothing was ever blacked out to protect Dennis's privacy.

Only the Crown Prince and King of Tonga held out. My first five visits, I'd regularly dropped letters and gifts to His Majesty and Tupouto'a, and they had regally ignored me. Now I asked Wayne to write a letter I could bring to Nuku'alofa. Wayne toned it down some for royalty.

Dear Sirs,

. . . For the last 25 years I believed and trusted my Government would follow through with justice and put this animal away from society, as the U.S.

State Department and the killer's parents, Mr. and Mrs. Sidney Priven,
promised to do in letters to the Tongan Prime Minister . . .

I am 71 Yrs old (almost) and my only goal in life (what's left) is to see
justice done. I need, no, I beg your help.

The morning after I dropped off Wayne's letter, the Crown Prince's secretary
told me to be at Tupouto'a's villa at three o'clock. The villa was the most magnifi-
cent building in the Kingdom, a winged Italianate eminence on a hill overlooking
arbors, pumpkin fields, and the lagoon. Tupouto'a met me with his chamberlain,
the nobleman Fielakepa, and led us into the piano room. He wore a cream-colored
suit and cream slippers, and sipped coffee from bone china.

"The father's letter, I found that very moving," he said with a humbled air. "Of
course I remember the case, ask me whatever you like."

"The chargé d'affaires said you wanted the case to disappear," I said.

"We wished that it had not happened here. The Tongan response was shame.
Shame and guilt that somehow this had happened because of Tonga."

"You told Flanegin the crime was unpremeditated—"

"That is true. But it did not matter what I said. I was expressing a personal
opinion. If I were to interfere, I would have had my head handed to me. The
police minister was very determined that the case deserved the death penalty. I
told Flanegin I could do nothing about the process. Why didn't he tell Washington that?"

Anything I wanted from the Tongan government, Tupouto'a said, Fielakepa
would provide. Fielakepa was the minister of lands and had fond memories of his
former teacher at Tonga High School, Deborah Gardner. The only thing the
Crown Prince couldn't deliver was an audience with his father. He said he didn't
have any pull there.

His Majesty was in his eighties and bent by age. I often saw him from afar, at
ceremonial events or in Centenary Church, but that was as close as I got. A couple
of times I talked to the King's private secretary, the nobleman Ma'afu, at the
Nuku'alofa Club, and Ma'afu would get me a beer and say, "Yes, but why now?
Why do you care?" then squint and stare off and nod his shaved head. Ma'afu was

from a long line of warriors. He was going to lay his body down before he'd let me near the King.

Still, Ma'afu did give me a letter from palace files that Peace Corps director John Dellenback had written to the King after Dennis's return. My government had blacked out the key passage in that letter. Tonga didn't.

Wayne's letters also worked in Western Australia. I showed up at the McMaths' modest house in Ballajura at ten on a Saturday morning. John was craggy and thoughtful, Marie was overflowing. They pointed out the green turtle shell on the wall Dennis had given them before they left Tonga, then we sat down in the living room and I started to describe this project. That was not necessary. I was saying that some people felt I was stirring up a mud puddle that had settled a long time ago when Marie interrupted me.

"This has never settled!" she cried. "I was a mother to Dennis, a big sister to him, John is the only one who knows about this. This is my ghost, I have to get this out—"

Deborah Gardner on the Tongan island of ʻEua during her Peace Corps training, December 1975. *(Victor Casale)*

The members of Tonga 16 were sworn in as Peace Corps volunteers in Nukuʻalofa on January 30, 1976. Deb Gardner took the oath along with Paul Zenker *(far left)* and Victor Casale *(left)*. Language teacher Tovo ʻUele is at right, in skirt and woven taʻovala. *(Mark Stiffler)*

Mike Basile, the assistant Peace Corps director in Tonga, had been a volunteer himself in Turkey from 1965 to 1967. *(Courtesy of Michael Basile)*

The King of Tonga walks around a potential oil field in Tonga in 1976. His Majesty had contracted with a Denver company to explore for oil. The scheme came to nothing. *(Robert Forbes)*

Wearing ceremonial dress, Emile Hons dances before the King in November 1975, the centenary of Tonga's constitution. *(Courtesy of Emile Hons)*

Frank Bevacqua was a 29-year-old volunteer from New York, restless, artistic, and moody, who fell in love with Deb Gardner in 1976.
(Courtesy of Mark Stiffler)

Dennis Priven, 24, a brilliant and volatile volunteer from New York who also developed feelings for Deb, is shown socializing at the house of the McMaths, Australian missionaries who lived near him in the village of Longolongo.
(Courtesy of John and Marie McMath)

Victor Casale, a 25-year-old New Yorker, was the most charismatic of the new volunteers.
(Courtesy of Victor Casale)

Volunteers in Ngele'ia on a Saturday morning, October 1975: *(left to right)* John Myers, Emile Hons (with bush knife), Dennis Priven, John Sheehan, and New Zealand volunteer Philip English. *(Courtesy of Philip English)*

'Akau'ola, the Tongan police minister, was enraged by detectives' bumbling in the case but was determined to hang Deb Gardner's killer. *(Courtesy of the Faletau family)*

Mary George *(center)*, director of the Peace Corps program in Tonga, at a picnic soon after her arrival in the Kingdom in March 1976. *(Courtesy of Michael Basile)*

High school teaching staffs were composed of Tongans and *pālangis*. Here is the staff of Tupou High, the leading high school in Tonga, in 1976. Principal Kalapoli Paongo is in the front row, third from the right, flanked by deputy principal John McMath and minister Vili Vailea. Third from the right, back row, is Paul Boucher, a member of Tonga 16. New Zealand volunteers Gay Roberts and Matthew Abel are second and third from the left in the top row. The Bird Lady stands at far right, beside Dennis Priven, both members of Tonga 14. *(Courtesy of Tupou High School)*

The primary school grounds, Ngele'ia. Emile Hons's tin-roofed house can be seen at left. Deb Gardner's house is in the foreground. *(Mark Stiffler)*

The white dress Deb was wearing the night she was killed, now stained with her blood. The dress remains in the Tongan police evidence file for the case. *(Courtesy of the Tonga Police Department)*

A grim Emile Hons, during the cleaning of Deb's house on Saturday morning, October 16. (*Courtesy of the Gardner family*)

The preliminary hearing in the murder case was conducted in the police court two weeks after Deb's death. The trial took place in the white clapboard building at left, December 13 to 23, 1976. (*Rick Nathanson*)

Deb Gardner's body leaves the Kingdom for Fiji en route to the United States, October 16, 1976. Uniformed Tonga High School students form lines to see her go. Mary George is at right, back turned. Francis Lundy is at left, in white shorts. (*Courtesy of the Gardner family*)

"What do you say when someone you are close to is gone completely? No more exuberance, vivacity, affection, considerateness. No more 'sparkle.'" From Mike Braisted's journal entry the day after Deb's death. (*Frank Bevacqua*)

22

At Sea

Laura was frantic. The police were planning to move Dennis back to the prison farm. Hu'atolitoli looked like an internment camp, with raw cinderblock buildings and concertina wire fences on which the prisoners hung their washing to dry. If the Tongans moved him to Hu'atolitoli, Dennis would kill himself.

Laura went to Mike Basile. Mary had left Tonga for the month of November, to attend a NANEAP conference and make a trip home, so Mike was now the highest American official in the Kingdom, and Mike went to see Dennis. It was late in the day. The little wooden cell was dark but for the high eastern window, and Dennis squinted when the door opened, and Mike was not prepared for how fearful it felt.

Thankfully the door stayed open. "Dennis, I hear there's a problem with your being moved to Hu'atolitoli."

"If I am not here, I will kill myself."

"What's the problem with being out there?"

"I'll die there. It's a terrible place. People get killed there and no one knows it. I have been in Hu'atolitoli already. The men there are truly crazy, they will kill me. I want to be here. If I'm not here I will kill myself."

"Well, hold on—don't do anything. Let me see what I can do."

Dennis looked at Mike for the first time. "You will keep me here," he said evenly.

"I'll do what I can," Mike said formally.

Mike couldn't raise the Crown solicitor by telephone, and then Wyler said that Tevita Tupou was to be found at the Nuku'alofa Club at that hour. So Mike drove downtown and parked across from the palace.

The club stood catercorner to the palace, a handsome square building with a wide verandah. Formerly the members' names were Zuckerschwenk and Fatafehi, noblemen and German traders. Now the membership had come to include Tongan professionals, so long as they were male, of course. The club was dignified but not crusty, who could maintain a crust in Tonga. There was a story that Prince Charles had wandered in one morning, clueless in Nuku'alofa, and the manager had asked if he was a member then told him to keep moving, and the experience had deeply pleased the prince. At last, to be tossed out of a club!

Mike cracked open the screen door. The barman came over. "Do you have a card?" Mike scratched out a note for Tevita on a beer coaster.

There is nothing so soothing as Tongan twilight. The streets are still, the roosters have quieted. The pink-and-orange shout of the bougainvillea has died down, the breeze brings the smoke of cooking fires. But Mike Basile was upset. "You will keep me here." The order had taken him by surprise, also the insight Dennis possessed that Mike was attempting to stay neutral, and Dennis's determination that Mike should not be neutral. It was said that Dennis was insane, but Dennis had seemed coldly rational, aware of his interests.

Mike had liked Deb. She was free in her speech and her emotions, she was up-front, he could tell what she was feeling, and he had always enjoyed seeing her. Of course, she was sensual too. The love of life seemed to pour off her forehead and shoulders, and Mike had seen goodness in her every action. . . .

"Dear Mr. & Mrs. Gardner, I want to express my very deep appreciation for your daughter's service to the Peace Corps and to the people of Tonga. Know that wreaths of flowers were laid at our feet in her honor . . ."

He thought of his first year as a volunteer in Turkey. No electricity, no running water, no paved roads, and at night Turks came to his little house to talk about religion. He played cards all that year, while in Brazil his future wife was building schools, and the Peace Corps said to them, "What you do is on you, you are completely responsible for yourself." The only way Dennis could be held responsible was if he stayed alive.

The Crown solicitor was the best snooker player in the club, and it irritated him to break off his game, but Tevita brought Mike into the library and signaled for beers. He knew Mike, their wives played bridge in the Queen's game.

"What's the problem?"

"I have just visited Priven. I'd been told that he was threatening suicide, I wanted to verify it. He said he would take his own life if he is moved to Hu'atolitoli."

The Crown solicitor was silent. It was what all the *pālangis* said about the Tongans, they never let you know what they were thinking. At 34, Tevita Tupou was the promise of the next Tongan generation. He was smaller than most Tongans, more delicate, courtly, with a touch of London glamour evident in the muttonchop sideburns and tinted glasses, the psychedelic shirts he wore to parties. He had been adopted as a boy into a nobleman's family and groomed for leadership since, but he was a commoner.

"Mike, why are you doing this?"

"He is a Peace Corps volunteer."

"I don't understand, why all of you in Peace Corps are showing this kind of regard for a brutal murderer—"

"He's my charge."

"As Deborah Gardner was your charge." Tevita's face was flushed. "Do you know how he killed her? I will show you the pictures, it was worse than hacking an animal to death, and now all of you want him to be treated with kid gloves."

Mike said that he was now the highest American official in the land. The American government had a strong interest in keeping Dennis alive. If Dennis were removed to Hu'atolitoli he would commit suicide.

"Then let him. If that is what he says he is determined to do, why don't you let him?"

"Tevita, you know that I cannot do that."

"He is manipulating the system to his advantage."

"He is quite serious."

"Yes, I know, all of you in Peace Corps have such high regard for this serious and manipulative murderer, so all of you are now coddling and catering—"

"What do you mean, all of us?"

"It appears that the organization has given orders to volunteers to withhold testimony."

"There have been no such orders."

"Well, the police interviews have gone poorly. We are getting very little information."

"That is not policy. If you know of instances, you should inform me at once."

Silence. Mike had drained his beer in the shock, Tevita got to his feet. They had both retreated to formality, and Tevita said curtly that moving a prisoner to Hu'atolitoli was a police matter.

"None of us wants another dead American volunteer," Mike said.

"I will see what I can do . . ."

Tevita walked back to the snooker room, tried to begin another game. But his hands were trembling. Pink, blue, and brown balls fuzzed and danced on the felt, he could not see the game before him, and he turned and put his cue back in the rack.

The police brought Dennis out to Hu'atolitoli on the flatbed and stuck him in maximum, a small cinderblock building in one corner of the big yard, with just four cells in a row. Dennis curled up on the floor and ignored the food the guards brought him. When they shoved a cup of water in, he spilled it out on the apron of the cell. The police threatened to drag him out in the yard and hose him down if he did not eat. He ignored them.

The cell faced a grass yard with a roof of crisscrossing strings of barbed wire that made a crackled patina of the sky. At night a light was on in the yard, it was hard to sleep. The stone was dank, the cell had an inner and outer chamber, like a kind of internal organ, and he got through the Bible, moving methodically from page to page. It was possible to exchange a few words with the man in the next cell, even to peer at him. Prisoners had incised a draughts board in the cement apron so they could reach out through the bars and move checkers and play their neighbors. Not even proper checkers, but the red and white caps of toothpaste tubes, and the game was Tongan checkers. The pieces could move back and forth.

He was surrounded by glottal Tongan chatter, and woke up at night with nightmares, and the police said that he was acting crazy, that he sang strange songs in the middle of the night and took his shit in his own hands and threw it at the bars. A prisoners' chorus sang hymns, and a bell chimed the hour. For a bell the guards used an acetylene tank left by the Americans during the war, cut off halfway down and hung from a bar. The guards banged on it with a hammer.

The Bird Lady went to the central police station every day to find out whether 'Akau'ola had allowed Dennis to come back. But he had not, and she clung to the other people whose lives now turned around Dennis. Paul Zenker came over to Longolongo to fix her bicycle. He remounted the chain guard but kept losing the screws in the gravel. Paul was in an awkward position. He had liked Deb, too. But on the night of her death he had said to other volunteers at 'Atele, "So where is Dennis?" and they had gotten angry at him for even caring. So he visited Dennis, but he did not advertise it to his friends. Then it was seven-thirty and he was going to a movie with the Bouchers, and the Bird Lady went along, and rode Paul's bike so that her skirt wouldn't catch in the open chain, and after the movie Paul finished fixing the guard.

The next day in church the minister gave a sermon about how Christ had fulfilled the Law by extending it to the spirit as well as the body, then the minister began talking about the Ten Commandments and started right out with the sixth and Dennis, and to her surprise the Bird Lady was so upset she began trembling on her mat. The minister imagined the scene that night, Dennis sharpening his knife and going over to Debbie's, planning to kill her all the while.

The Bird Lady wanted to stand up and thunder to all, Do you *know* that Dennis is the person who killed Debbie? How do you know? And if he did, do you *know* how he did it?

She did not go to the feast after the service. She would have been afforded an opportunity to speak, and she would have gotten angry. Instead she rode off to school to sit in the staff room and write letters.

"I wish there were some wind. It's stuffy in here even with all the windows open."

That night she rode home by way of the Bouchers and the Bouchers asked her to stay for supper. They ate and listened to the *Peter Pan* soundtrack.

It might be miles beyond the moon or right there where you stand—
Just keep an open mind, and suddenly you'll find, Never never land.

The next day everything changed. The Bird Lady stopped off at the jail after school, and an officer said to wait, the police would have something good for her, and a yellow van came up, and there was Dennis, sliding out pale and weak. She greeted him overjoyed, and he smiled back, and the officer even chuckled. They let her sit with him for a while. He had gone five days without water, and the Bird Lady scolded him for taking such a risk.

The other friends came in for Dennis's homecoming, and there was a new set of rules. The friends could sit with the prisoner in the charge room, but Dennis must not eat there, the charge room was not a restaurant, the prisoner was not to become a distraction. It had been reported to 'Akau'ola that Dennis had played frisbee in the yard of the station, and come and gone to the bathroom on his own.

Laura set out to coordinate all their visits.

"You must only visit him a half hour tomorrow to minimize the disturbance," she said to the Bird Lady.

The Bird Lady promised, then wondered if she would keep her promise. What if Dennis wanted to talk? That was her role, to listen and keep his spirits up in the weeks before the trial. "We've been supporting him some & also letting him talk safely," she told her parents. The Tongans did not understand. They seemed to think that talking with him for hours made his life too easy. When she had tried to explain to a neighbor about mental hospitals back in the States, the woman said, "In America you do not consider murder a serious thing."

"I hope Dennis can get back to the U.S.," the Bird Lady told her parents. "I hope it, I hope it, I hope! And as soon as possible also. It has seemed to me that the picture of himself he's projecting won't hold up forever & somebody who can help should be there when it doesn't."

The police posted a list of approved visitors. Mary wasn't on it, Dennis didn't like her proselytizing him. And neither was Emile. Dennis's old friend had come back from the States and visited him once, but that was it for Emile. He was never going back to the jail. Some things Dennis said had left him chilled.

There were just six visitors. The Pauls and Laura were on the list, of course, and so was the Bird Lady. And John McMath and Vickie Redpath. The teacher of Shakespeare and romantic poetry who had wanted to bring Deb's body home, Vickie was also close to Dennis.

So Dennis had built a new family in Tonga, a better family: Laura and Paul were the parents, Paul Zenker was the new brother. With an unexpected sister thrown in, the Bird Lady.

He was estranged from his Sheepshead Bay family, though the Peace Corps acted as go-between. HE DOES NOT RPT NOT WANT HIS PARENTS TO ATTEND TRIAL. AND NOT JAY EITHER. His family blew him a kiss back. MOTHER SAYS FAMILY SENDS SUPPORT AND CONCERN FOR HIM IN THEIR THOUGHTS . . .

Mimi was draped over the kitchen table. One Peace Corps memo described her as "catatonic," and her close friends, trying to protect her, told other women in the co-op building that it wasn't Mimi's Dennis. That was hopeless. Later one of the close friends whispered to the others that it was true after all, it was Sidney and Mimi's Dennis.

The neighbors wondered why the Privens weren't going to Tonga. They were middle-class people, Sidney was a printer. Couldn't they scrape up the cost of a ticket? Mimi told one friend they couldn't afford it.

Do you want us to send Dennis a cable telling him to write his parents? Peace Corps asked. No, the Privens said.

The family held out hope that Mary was right, that it was a frame-up. "Is a Tongan responsible for the death of Miss Gardner? And is there a conspiracy to blame Dennis for the death?" they wrote to Clive Edwards in Auckland. Clive said that he was looking into the question. He was paying two secret agents back in Nuku'alofa to be his eyes and ears. It was possible that another man with whom Deb had been involved was responsible, a man called Frank Bevacqua.

But he'd heard nothing yet to support this view. The police had interrogated Frank for hours, and in fact Clive had evidence to the contrary.

"Your son has not made any admission to myself, or to the police, but Tom Finau, a local lawyer in Tonga, had interviewed him prior to my arrival and he says that your son advised him in a vague fashion that he was involved in a struggle with Miss Gardner and was there at the time."

In early November the Hawaiian psychiatrist filed a report on Dennis, and Clive could no longer sidetrack the prosecution. He sent the report on to the Crown solicitor and the Crown prosecutor, who were surprised to discover that Dennis had been diagnosed as a latent paranoid schizophrenic. The Crown had spent weeks shoring up its case that Dennis had been in Ngele'ia that night. Now it appeared that Clive was going to mount the first insanity defense ever tried in Tonga.

Neither prosecutor believed that Dennis was insane. And though it was difficult to destroy an expert's evidence without another expert, and there was no psychiatrist in the entire Kingdom, still the prosecutors felt that if they laid out the facts of Dennis's days leading up to the murder, they could demonstrate that he was not insane. He was a highly intelligent chemistry teacher who had seemed normal to his students. He had bicycled across town to her house carrying several deadly implements and afterward said *"Mālō e lelei, nofo ā"* to a Tongan sitting under a mango tree as he bicycled away.

A crazy person did not say, "Good day to you, and goodbye."

He had divided his earthly possessions among his friends, then turned himself in. His suicide attempt had been feeble. At the station he had warned Kuli about the poison.

Their problem was that the prosecutors knew nothing about Dennis's social landscape. There had to be some background to the murder, a story. They believed that Deb had rejected Dennis and he had been possessed by jealousy. But interviews with *pālangis* had come back with gaps.

The Tongan officials were not aware of the parallel American investigation, which had collected a lot of information in Washington.

When he had roused Emile at six in the morning in Seattle, the associate counsel had gotten a description from him of the relationship between Dennis and Deb.

"Dennis liked Debbie very much but she did not reciprocate his affection although she was polite to him. Debbie's unresponsiveness was a great source of frustration to him. The Monday before the murder, he dated Debbie and was quite elated that she seemed to have been more affectionate toward him than usual. On Tuesday, however, she turned him down for a date."

The official Peace Corps record on Deb Gardner's service suggested that her turning Dennis down was the reason he had killed her. For cause of death, the record said, "Stabbed by another male PCV when she rejected his repeated advances. ETOH use [alcohol use] by assailant reported." Nothing about paranoid schizophrenia.

Then there was the cyanide. The cyanide was an open secret among volunteers in the days after the murder. None of them had to have the bottles tested in New Zealand. Emile had told the associate counsel when he called him in Seattle that morning that there had been a bottle of cyanide left in Deb's house. Gay Roberts had learned that the label on the bottle of liquid said HCN for hydrogen cyanide, apparently in Dennis's handwriting, and that the powder was a form of cyanide too, with a label saying, "Poisonous if liberated with acid."

Gay had told the Bird Lady, and the Bird Lady had even had a conversation with Dennis about the cyanide.

Generally Dennis did not talk to her about the case. He did that with the Pauls. But he hinted to her that he needed to talk to her, too, about the murder. "I just hope for some wisdom if he does tell me much of the realities of that confused & confusing night," she told her parents.

Finally the conversation happened. As police wandered around them, Dennis brought the facts up in indirect fashion. They had to talk in unclear sentences, with elliptical references and pronouns.

"There's a rumor going round that the intention was to kill her with a chemical," he said. "It is necessary to refute that rumor."

It took a while but the Bird Lady caught his meaning. Dennis felt misrepresented by the suggestion that the cyanide had been intended for Deb. The cyanide had been intended for himself. The Bird Lady framed the thought back to Dennis as a scientific hypothetical.

"If he had intended to kill her with the cyanide, then that is how she would have died."

Dennis gave the neat if-then statement some thought before he nodded and smiled.

The dark cell could be a horror chamber at night. The policemen screamed at a thief in cell 2, then beat him against the walls, demanding to know where the bicycle was, and for hours after, Dennis heard him moaning and was thankful for the privileges that came with being a *pālangi*. The police grew relaxed around him. There were to be no visits on Sunday, but the Bird Lady dropped by anyway to leave a note, and a policewoman put down her romance novel and brought Dennis out.

"It's against the visiting rules to see you, I don't want to break the rules," the Bird Lady said.

"Don't knock it, I'm happy to get out of that cell."

Dennis rambled on about street lights in Brooklyn, or the life and games of Avenue X and Avenue Z, and the Bird Lady gazed at the shabby charge room walls with their chipped paint, memorizing the different layers. Sometimes he told the Bird Lady how the case would go in the United States, but she felt he was deluding himself. "Oh, he wouldn't be in the jail now, I think he's right there," she told her parents. "He'd be, in all probability, in a mental hospital, with the door rather firmly shut behind him."

One Saturday in the market, the Bird Lady bought a large triton shell for Gay and then a basket seller asked the Bird Lady about the trial of the boy who had stabbed the girl in Ngeleʻia.

"But we don't know that until after the trial, do we?" she said brightly.

The man retreated behind a pile of baskets and grinned, then said, "You *pālangis* say the funniest things!"

The Bird Lady agreed, and left.

She knew more than she had told the basket seller. Over the weeks, Dennis had been talking to the Pauls about that night, explaining it and maybe even understanding it. That thought made her glad. She wanted Dennis to understand what happened. He did not tell her very much, he protected her from that sort of information, still through one conversation and another, the Bird Lady had gained some insight into his actions.

By killing Deb, Dennis felt that he was performing a service to society. Deb had led several men on and disappointed them. She was insincere and deceptive and had abused the power that her looks gave her.

But Dennis recognized that society disapproved of what he was doing, so he would accept society's price. By killing himself he would prove to the world that he had done what he had done to Deb out of a generous and not selfish motivation, he was willing to die for the good that came out of the murder. It was also practical because that way he could control his fate, and the last thing Dennis ever wanted was to be in other people's control.

The Bird Lady hypothesized that Dennis's plan had been a surgical one, a neat one. Maybe he intended to render her quickly unconscious before killing her, maybe he had brought the pipe to perform that task. But very quickly the plan had gone awry. Deb had gone out of his control, and probably Dennis had been surprised himself by his own reaction, how messy and berserk the situation had become, and how quickly.

"You normally do things well," she had ventured with him once.

"But—" He held up his wrists, as if to say, I botched that.

"Maybe that's because you didn't really want to die."

That annoyed him, he shook his head and denied it. But another time, he said of his suicide efforts, "The dead end seemed a little deader."

Dennis's friends may have helped him toward understanding, but they had been placed in a curious position. There is no legal privilege for friends, the police had a right to know these things. The police had interviewed Paul Boucher, but they never did try and interview the Bird Lady, and the Bird Lady could justify not going to the police because she was doing all she could to make sure that Dennis got the help he needed.

Dennis could only get that help back in the States. So there was something extraterritorial about Dennis's new family; they were Americans who knew better than the Tongan community how the matter should be handled. But what commitment had Dennis made to the friends? The friends were isolated from other volunteers now not just because they liked Dennis, but because they may have known things. In a certain sense, anyone to whom Dennis had given incriminating information and who did not help the police as they flailed had been made into a kind of accessory.

Outside the Kingdom, the story had died. After the Election Day announcement there had been no follow-up in the newspapers. Reporters in Seattle-Tacoma and New York had given up on talking to the families, and in Washington the White House transition was absorbing everyone's attention. The press corps was interested in Henry Kissinger's memoirs, Jimmy Carter's Georgia friends.

When Bill Kinsey had been tried for murder ten years before, many African reporters had covered the trial, and they were joined by several journalists from out of the country. This time around only the Tonga *Chronicle*, which some expats called the Minute of Silence, because that is how long it took to read, had stayed on the case, covering the story every other week or so, and only one *pālangi* journalist outside the Kingdom had any real knowledge of the matter. O'Loughlin, the AP bureau chief in Sydney, gave some thought to going back out to Nuku'alofa for the trial, but none of his American clients asked him to, not even papers in New York and Washington state, and though he might have assigned himself, he was not particularly inclined to spend two weeks in a sleepy Pacific town as Christmas approached.

Besides, he had a reliable stringer in Paua Manu'atu, a shy and thoughtful man with a craggy brow and noble blood, who had been editing the *Chronicle* for many years. Paua had done a journalism fellowship of several months in Hawaii, but by and large relied on Peace Corps volunteers.

Sitting in their office at Tungī Arcade with coconuts on the desk alongside the manual typewriter, Paua and Rick Nathanson divvied up responsibilities for the trial. Rick would spend as much time as he could in the courthouse. Paua would drop in when he had time. At the end of the day Rick would file a story, or they'd put their notes together, and Paua would telex Sydney with a short article, which O'Loughlin would wire to New York.

Radio A3Z was planning to cover the case too. From time to time its reporter would go to the courthouse. But for all intents and purposes, there was just one reporter on the case, wild-haired Rick Nathanson, wearing cut-off denim shorts and flip-flops and taking notes on a government payout of 2 pa'anga a day.

In early December he went to the Supreme Court to cover the pleading, and afterward Mary took him aside.

"We don't think the murder is the sort of story that the Peace Corps ought to be focusing on."

"But Mary, it's my job, it's an important story."

"This story is not good for anybody," Mary said. "The in-country volunteers, the program, or Washington."

On December 8, Dennis said the only words he was to say throughout the trial, Not guilty. His trial would begin in five days.

The McMaths were to miss it. They left the country the next day.

They had shipped their things and euthanized Brutus of the fishhook tail. They couldn't bring Brutus back to Perth, and weren't about to leave him to the tender mercies of their Tongan neighbors, who had already put the McMaths' pet billygoat into an 'umu one Sunday for the feast after church. So John gave Brutus five injections of Nembutal then buried him in the backyard.

Then they went to the jail.

It was Marie's first visit. Tongan summer now, a stinking hot day. John stepped inside. Marie waited outside the doorway with the two bikes. She did not want to say anything to Dennis, only wanted to see him. She was anxious, and she let her eyes wander about the station. It surprised her how open the place was. She could not see any locks or bars. But then where was Dennis going to run to?

She had last seen him two months before, on the day of the murder. The noise of the door had wakened her, she saw Dennis walking down the road with his laundry on his back. Marie sat up in bed and something was different. Then she felt it brushing her face. She touched her hair and more of it came away in her hands, clumped in her lap. She ran to the bathroom and the front part of her hair was gone, her shining beautiful hair.

Then she ran to Dwayne, he was asleep but he was fine, and locked the door and started to cry. She had heard nothing, she hadn't wakened. Brutus had not even growled, Brutus knew Dennis.

She gathered the hair and was ashamed, she got a piece of paper and swept all the hair onto it and put it on the dresser, waiting for John. Now this is what it is like to be raped, she thought. Then she cried harder and knew what the Jewish women felt in the war when Nazis cut their hair in a sign of ownership. Why do you think the Germans took such pride in cutting the Jewish women's hair? The power, the invasion. She had been naive and stupid, she had been invaded.

She studied the hair. It was jagged, not cut in a straight line like a scissors would have left but as if he had held the hair in one hand and used that knife in the other, the knife she had commanded him never again to take out in her house, and Marie could not understand why she had not woken up. He must have looped her hair about the blade of his knife with one hand, then held the hair and cut with the other.

"Could it have been anyone else?" John said.

"No, it could not have been anyone else. Brutus would bark if it was anyone else."

Brutus was a *pālangi* dog, he had one bark for *pālangis* and one for Tongans.

"Never is he coming to this house again," Marie said, "and you are going to deliver the news."

"I will talk to Dennis tomorrow, take him aside at school," John promised. He'd be firm but goodnatured. "Unfortunately, mate, the door's no longer open for you. . . ."

Marie dropped the hair in the bin and the next morning when the volunteers came to the fence with the news from Ngele'ia, she buried her face in her husband's neck and said over and over, "It could have been me." Later, thinking about it, Marie said, "He was sharpening his knife on me," while John being more logical said that Dennis had been rehearsing, seeing how composed he could be with his knife in the presence of a recumbent woman. Divine intervention had kept her from waking up and suffering worse, the McMaths both felt that to be true.

Then there were rumors. People had seen the look in the McMaths' faces. Something had happened to make Marie lock the house up. A detective came to the house, but Marie didn't open the door.

"I've got an ill father at home, I am going home," she said.

"I can make you come to court and talk about Dennis Priven. I can subpoena you to come."

"You can, but you can't make me talk. I'm not going to speak against a man when I don't know why, what the angle is."

Now they were leaving. Marie was traumatized and sad, and did not see what had happened to her as being evidence bearing on the murder. Her children were looking forward to seeing their ailing grandfather at Christmas.

A door opened at the side of the charge room, and John stepped out and Marie saw Dennis, and Dennis saw Marie and dropped his head, looked away. He could see what she was feeling. Marie saw that in his face. He knew that Marie knew just what he had done to her.

Tevita and Christine had worked on the case for more than a month, and then Tevita took the Saturday before the trial off to go fishing and try to distract himself.

A few men set out in a small boat when it was dark, they shivered against the dawn. When the sun came up the boat started up and down the swells of the sea, and Metui said, "Here is a school of tuna, put out your hand lines." A tall bearded Peace Corps volunteer was in the party, he landed a five-pound bonita, and Metui filleted it and held out the first piece on his knife to the volunteer, but Brandewie hesitated till the banker, Bohart, said, "The Japanese pay $10 a pound for this stuff and it ain't near this fresh," and the men made short work of the fish and Metui flung its skeleton into the water.

He took them to an atoll two hours out. They fished all morning and had lunch on the atoll, and Metui, low-slung and bull-chested, told the men about sharks. When he had started out diving, he had been as afraid as anyone of sharks and did all he could to avoid them. Then one day a big shark had come toward him and Metui believed he was about to die and he did the only thing he could think of and swam at the shark with his speargun, and the shark went away.

So that is how he learned to hunt shark. The shark was as afraid of Metui as Metui of the shark, and he had not encountered a shark yet he needed to swim away from.

Then they had sixty pounds of fish in the little boat, jacks, groupers and parrotfish, and it was mid-afternoon and Metui dove and was gone for a long time. The three others became concerned. Maybe Metui had finally met the shark who was not afraid of him.

But he broke the surface with a gasp and didn't have his spear, just the end of the rope. "Give me your spears—"

Again he was gone for a while then came up with the rope ends and told the men to pull on them, he was going down again.

This time when Metui rose a great shadow was under him, it looked to Brandewie as if he had found a Chevy down there. Metui held the gunwhale with both hands, breathless. "*Helepelu*—" Tevita gave him the machete.

Now a splash, and a large ray boiled the sea. Four spears stuck out in the air from one wing, and its tail swung in agony from side to side, big as a man's arm and barbed, till Metui held the machete against the base of the tail and the tail cut itself off.

The men in the boat didn't like the idea of bringing the fish in. There was no room for this fish, they were overloaded already, the sea wasn't calm. But when Metui flopped one wing on the gunwhale, the men started pulling it in, slowly, to keep the boat from swamping.

Then it started to convulse. The wide fish crunched itself like a mottled fist, and relaxed for an instant before it started again.

Metui called out for the men to push it back into the water, and he hung with his elbow over the gunwhale.

The first one came out into the sea looking like a crumpled handkerchief before it calmly opened and glided away from the chaos. Another came, and another. They were the size of dinner plates, a foot across and light in color, not black like the mother. They floated easily away in the sea, flattening out. No one said a word, no one cared for himself. By some supreme chance they had been included at the doorway where death and life exchanged places.

Then the contractions were over, eleven or twelve babies had gone into the sea, the ray was dead, and the men pulled it the rest of the way in.

They were wet and exhausted. But before long they were able to see the lights of Nuku'alofa and hang their hopes on it between times that the swell rose and blocked the city off.

Tevita thought about duty and secrets. The fish had done all she could and the men had honored her by not taking her offspring too. The two *pālangis* had shown the mother respect, they had seemed humbled, the volunteer in particular had been solemn, awed.

He looked over at him now. "Did you know Debbie Gardner?"

Brandewie nodded. "There was a party at the Dateline a few days before she died. She and my wife and I walked along the pier together, I remember she was giddy and outgoing, and she was gorgeous."

"But your organization. It has no compassion for her—"

"Well—the director says that he didn't do it."

"Yes, the director is a buffoon. Now they are spending tens of thousands of dollars to save him, and the psychiatrist from Hawaii is saying that he is crazy."

"Why don't you get a psychiatrist of your own?"

"Of course, and who is going to pay for it? The government is flat on its assbone. You'd think the Americans would pay, they are paying a fortune to bring Neti—"

The boat came into the wharf and people were hanging around, wondering where they had been, and Metui was the last to hop ashore, and the story of the ray was his to tell. He had seen her nosing around the coral head in an odd way, not really trying to escape him, then later he understood she was looking for a place to deliver her babies so that they could find refuge in the atoll. They were there now.

23

Tonga in the Dock

P lease state your name and age."

"Mary Evelyn George. Forty-five years old. Excuse me, forty-six."

"Did you know the deceased, Deborah Gardner?"

"Yes, she was a member of our organization."

"And when did you see her last?"

"On the afternoon of October 14, 1976."

"Where was that?"

"At my office at the Peace Corps."

"Was she in good health?"

"She appeared to be in very good health."

"When was the last time that you saw her alive?"

"That was the last time I saw her alive."

"And when is the next time that you saw her?"

"The same evening, at Vaiola Hospital. She was covered by her sheet. All but her feet."

The first witness sat in a box with her back to the gallery. A jury of seven farmers was on her right, and Tevita stood before her in a black robe and a white horsehair wig that didn't quite cover his hair.

The courthouse was simple and dignified, a large room in the manner of a Pacific longhouse with open doors at each of the four corners. Casuarina trees with their long gray-green needles shaded the doors closest to the road, and fans near the open windows battled the still air. A great 'ovava tree made a sort of second court outside.

Dennis sat in the dock. From time to time he looked out at someone he knew in the courtroom and smiled privately through the heavy beard. But mostly he gazed ahead impassively, perplexing and terrifying the schoolchildren who had come to get a look at a murderer. Laura had brought him his second pair of glasses that morning, the first being broken and held in evidence, but the police said she would have to fetch them back at the end of the day. The prisoner was not permitted to have glasses in his cell.

"He's looking pretty good in his blue shirt & dress pants. They haven't let him shave at all, so his beard is big. The judge at his high bench wears a red and black robe. He looks somewhat like Santa Claus in his wig." The Bird Lady wrote down everything she saw, till a policeman came up to her with a summons to testify, and she had to leave. Clive didn't want her in the courtroom, didn't want anyone to think that Dennis had a loyal girlfriend.

Mary was clipped and brittle under Tevita's questioning, then it was Clive's turn and he slouched toward her, irresistible and heavy as lava. The courtroom belonged to him. He was more comfortable here than any of the others. The night before he'd been drinking at the Yacht Club on the beach road, and 'Epeli the palace secretary had asked him what his plan was, and Clive had looked up mildly. "I am going to eat Tevita Tupou alive."

Now he was determined to turn the court's attention to the case's interior reality.

"Can you describe your relationship with the accused?"

"Dennis and I initially had difficulty in communicating," Mary said. "He was due to end his two years at the end of the school year. He was so anxious to stay in the Peace Corps for another year."

"And could he stay?"

"There was a question of whether the principal wanted him, and that was very troublesome to him."

"Describe the accused's state of mind."

"He was worried about his future. There were obvious signs that he was depressed."

Mary was excused and sat with a clutch of friends at the front of the courtroom and behind the defense lawyers. There was Tanya Dobbs with her bobbed hair, and Mrs. Farquehar Matheson, a writer from California who lived in Vava'u, a slender woman ten years older than Mary, who'd come down to stay at Mary's house and give her support.

Dr. Puloka took the stand and was questioned for a long time. He described Deb's wounds in a dry and clinical manner. Number 10, number 11, number 12. Some of the smaller wounds were apparently defensive, parrying blows to the arms. "The blood flow ceases when the heart dies," Puloka said, then Lutui was called and sworn and said, Yes, that appeared to be Deb's dress, and that Dennis had seemed to him to be normal that night.

An older *pālangi* couple came through a back door. They sat down near Rick Nathanson and the young reporter kept glancing over at them, wondering how he might approach them during a break. But he was not sure what he would ask. He only felt sorry for the Gardners, he was afraid to go up to them.

Then the break came and they left and didn't come back, so Rick told himself that maybe he had not failed that test after all.

No, Deb's mother was home alone in Tacoma. What good would it do for Alice to go to Tonga? She wouldn't be able to take it, it would just have upset her more. Frank Bevacqua had sent her a mess of photographs of all Deb's friends but she didn't know who was who. She felt powerless. She was just going to study the newspaper, and clip the articles, and send them to Wayne, who was pinned down in Anchorage by a different set of feelings.

The parents did not know that Peace Corps's interest and Dennis's were now intertwined. Sitting in the courtroom, Mary slipped notes to Clive, suggesting one tack or another, and though the first lawyer from headquarters had warned her about that, that lawyer had wanted nothing more to do with Tonga when he left, and the lawyer who'd come out from the counsel's office for the trial, Barkow, a dark-haired good-looking affable man, was neither as detached nor as persnickety as Magid. Dennis was now DENNIS in the government cables, and if anyone had to mention Deb, she was usually "THE DECEASED" or "XPCV GARDNER."

Some years before, Sargent Shriver had testified on Capitol Hill about the support he got from the families of dead volunteers. "The seven families who lost a child in the Peace Corps are among the most ardent boosters of the Peace Corps," he said, and after Peverley Kinsey died on the picnic rock in 1966, Peace Corps flew her mother out to Tanzania for the trial, along with her sister to give her company. Now officials at the building across from the White House were struck by how polite Alice Gardner seemed, and were quietly thankful that the family seemed to be so accepting. The Gardners are not seeking revenge, the staff reported to one another, they are impeccably gracious. And though Peace Corps was in frequent touch with the Privens, giving them the latest developments in the case, it was telling Alice Gardner nothing.

Alice was dependent on Rick for everything she was to learn about the case, and Mary seemed determined to minimize that coverage. Mary had approached Rick that morning outside the courthouse.

"Rick, it's not a good idea to cover this. There are much better things to write about on the island."

"You know what, Mary, you're right, I should look for happier stories."

And Mary went into the courthouse, and a minute or two later, Rick went in too.

The first volunteer witness to be called in the trial was Victoria Redpath. The prosecution needed to establish that Deb had been alive at nine-thirty that night after weaving class, and then Tevita had asked her about the social background to the case, and Vickie as was her wont was plainspoken.

"On occasions Dennis and I discussed his relationship with Deb. He was interested in her maybe as a potential girlfriend. My feeling was that he should pursue friendship rather than the potential of being her boyfriend. Deb was interested in another man at that time.

"I'm not sure how he took my advice. He may have found it difficult to think of a friendship only."

Vickie even told the court about the October 9 party at the Dateline when Deb got drunk and Emile walked out with her and Dennis got angry.

"Dennis told me that Emile had taken advantage of Debbie. Dennis was annoyed with Emile. I had never seen him so unhappy. He was confused about where his friends stood. He was feeling distrust."

But when Clive got up to cross-examine Vickie, he tried to defuse any talk of a triangle.

"Was she seeing anyone in October?"

"In October I don't think she was interested in anyone in particular. She was enjoying freedom and was interested in several people."

But as for Dennis, yes, he was depressed.

"I was worried about him. I told him he should leave the country. He was disappointed. He appeared to be withdrawn."

Then Vickie was excused and took a seat beside Laura Boucher.

Vickie had been close to Deb. They'd hung out together, stuffed notes in the cracks of each other's doors. "Sorry that I didn't make it over, Deb, I got caught up making that batch of spaghetti. . . . So Debbie—will be over this weekend or if you have a second stop over. Vic."

Then the murder had happened, and seemed to transform Vickie. Early on she'd disliked Mary, said that Mary was objectionable and high-handed. Now she was more respectful, thankful for Mary's support.

The ordeal had brought Vickie close to Dennis, too, had brought out in her those feelings that Dennis pulled from people, devotion, fear. "Debbie is dead, we have to worry about Dennis," Vickie had said to her neighbor, John Myers, and Myers said to himself, "To hell with that, a person's never dead till the last person that loved them died, we shouldn't forget Debbie." And it broke Myers's heart that the Peace Corps had lined up behind Dennis.

But he did not go to the trial. Deb's friends from her own group weren't there either. It was summer break for schools. Brombach had gone to Fiji, Stiffler and Lindborg had gone to New Zealand. They anguished over the fact that they'd miss the trial, but figured there was nothing they could do about it, and Deb would want them to follow through on their vacation plans. As for Judy Chovan, when her parents had found out about the murder, they sent her a ticket to come back to Akron for the holidays.

For all her popularity, Deb did not have close friends in the way that Dennis did, did not inspire the same loyal feeling. She was something of a loner by nature, closest to her brother. She wasn't standoffish, but she was independent. No one got superclose. She enjoyed having many acquaintances, she separated herself from groups.

And the men who had adored Deb? Mike Braisted was getting ready to propose to a woman, Emile was lying in the grass outside the courtroom, waiting to be a witness. Both men saw Dennis as sick, felt he needed to be hospitalized.

Frank Bevacqua did not share that view; he thought Dennis was calculating and brilliant. Frank followed Vickie to the stand as a prosecution witness. The Crown called Frank to testify that he had last seen Deb at 6 P.M. and then Dennis at 9 P.M. or so.

Yes, he had been Deb's boyfriend in 1976, but it had ended in August.

Then Clive rose and asked Frank one thing.

"After you and Miss Gardner split up, did she have a permanent boyfriend?"

"No. She had no steady boyfriend."

Frank was excused, taking a seat at the back, and murmured angrily to his new girlfriend Jen from Australia, Jen who had dated Mike before, that the Peace Corps position was a farce.

The Tongans in the courthouse shared this feeling. Dennis was guilty, why was the American government fighting for him? Where was the directoress before this had happened? She had equal responsibility for both these young people, did she not?

Then the witnesses from Ngeleʻia began coming to the stand, and the Tongans watched in amazement, and fear, as Clive tried to shift the blame.

The attack was two-pronged. Any witness who had known Dennis, Clive would use to say that Dennis was losing his mind. Any witness who claimed to have seen him in Ngeleʻia, Clive would cut to ribbons. Oti Tevi was sitting under a banyan tree that night when a *pālangi* bicycled rapidly away from the primary school.

"*Mālō e lelei.*" To which the *pālangi* in the dock had responded with a natural greeting and farewell. Perfectly normal.

Clive undid the farmer with a few strokes, and even hinted that he was drunk.

"And who were the other two men you saw that night, who came looking for a phone?"

"I don't know the other two."

"Yet you are sure you saw the accused—"

"Because there was a light on at Savieti's home."

"Had you ever seen him before?"

"Yes. About every week."

"The previous week?"

"I don't remember."

"The week before that?"

"I don't remember."

"Every week. . . . Did you speak to this *pālangi* before?"

"Not before that night."

"How many *pālangis* come to the primary school?"

"There are a lot of *pālangis* who come."

"Yes. A lot."

Le'ota Naufahu had seen Dennis standing in Deb's door. She was 26, a mother, friendly, even sweet, and at first Clive was respectful, nodding patiently as Le'ota spoke, before he began to paint her as a busybody who had inserted herself into a drama.

"Now what were the dates you'd seen him before?"

"I have seen him many times, I can't give you dates."

"The times then."

"I can't give you the times."

"September?"

"I can't say."

"June?"

Le'ota was unprepared for grilling, she merely shook her head.

"Well then when did you see him?"

"He came with another Peace Corps volunteer who lived at Ngele'ia."

"That night you saw him standing outside—was it light at that hour?"

"No! It was dark, it was night!"

"But his clothes, were they light?"

"The bottom part of his clothing was dark. I can't recall the upper part."

"You can't recall the upper part. Was he wearing glasses?"

"I didn't notice."

"So you didn't know who it was, did you?"

Clive had warned some of the Tongan witnesses beforehand he might have to be sharp with them, that is what a lawyer did. He had seemed fatherly when he had said that, kind, sitting there with his square leather wigbox. Now here they were in his courtroom and all kindness was gone, and soon Le'ota became flustered.

"At the lower court in October you never said that you had seen the knife."

"The police did not ask me," Le'ota said. "But I did."

"And when you saw the knife, where were you standing?"

"I was standing in the road. But I had a clear view of it."

"And in the lower court you were asked if you had seen the accused once before and do you remember what you said?" Clive held up the record from the preliminary hearing. "You said no."

"Because I didn't see him once before."

"I have no further questions."

Tevita jumped up. Le'ota was vital, the only adult who had seen Dennis at the crime scene.

"Were you shown the knife at the preliminary inquiry?"

"No, I was not."

"Was there light on the accused when you saw him?"

"The light was on in her house. It shone outside the house."

"Why did you say that you had not seen him once before?"

"Because I had seen him on several occasions. They asked if I had seen him only once. It was not just once."

Le'ota was excused, and gave Clive a dark look leaving the courtroom. She had not grasped what was being done to her until too late, and now she felt embarrassed.

Tevita had a good house on the east side of town, a *pālangi* house, and that night when he pulled into the garage and closed his car door, he found that he was afraid to step into his own backyard.

He had an apprehension he was to have again and again during the trial, that an American agent was lurking in the bushes to knock him over the head. The Americans had many operatives overseas, they crossed borders at will.

Now some man from the CIA is waiting for me—

But he bestirred himself. No, that is nonsense, that is not logical. And having convinced himself, he went up to his house.

Many Tongans shared Tevita's suspicion. Some police felt that a *pālangi* stowaway who'd been taken off a boat from Samoa and put in cell 2 was a CIA operative, there to communicate with Dennis. Since when did *pālangis* ever stow away from Samoa? One *pālangi* after another had come to the International Dateline Hotel, all on Dennis's side, and now there was an American agenda to blame a Tongan for the murder. Mary George had whispered that Dennis was falsely accused, the *Fiji Times* had reported that a Tongan youth had killed Deb, now a rumor was going around that Deb had had a Tongan lover, and Clive Edwards the greatest lawyer the Kingdom had ever produced was getting $10,000 from the Americans to suggest that Dennis may have been framed.

It was also reported that Peace Corps had booked a seat for Dennis on every flight leaving the Kingdom, betting that Tonga would soon want to wash its hands of the mess. This was a bad bet. The Tongans did not want to wash their hands of it. The American agenda frightened them in an instinctive nationalistic way. If they washed their hands, the stain would never come off Tonga, it would stay and stay, they were cannibals who had tried to frame a *pālangi*.

"The Crown calls To'a Malanga Pasa—"

The boy was brought in by the police from under the *'ovava* tree across the way at eleven in the morning on the second day. Tall and calm, wearing a dark skirt and an unblinking expression, To'a had been preparing himself for weeks, in meetings with police, in prayer. But he was not prepared for the crowd in the courtroom and fumbled when asked his name, and his voice was thin as Tevita began to question him. He did so in Tongan, and the process was drawn out as the boy's words were translated into English.

"I live at Ngele'ia. Face the primary school.

"There are homes within the schoolgrounds, where *pālangis* live. I knew those *pālangis*. On the right was a girl, Tepola."

Then it was October 14, and To'a spoke as if he had rehearsed, and maybe he had. Still, the events were so fresh.

"He put her down face-first, with her head near the door. I saw the girl leaning up on her hands. The girl was Tepola. I had seen her often before.

"Yes, that is the man there. I had seen him several times before."

Tevita sat down, and then Clive got up and began mildly enough. After all, the boy was just a boy.

"You said that you left the prayer meeting and went to your parents' house to read a book."

"Yes."

"You were at your parents' house for ten minutes. Did you read a book and also wash in that time?"

"It was a comic book."

"A comic book." Clive nodded. "But you say that you were going to study that night."

"I was preparing for the New Zealand school certificate exams."

"The school certificate. What was the comic about?"

"Cowboys."

"Cowboys and Indians. I imagine there was shooting and fighting in the comic?"

"Yes."

Clive went back to his table, and something had been done to the boy, though To'a was not conscious of it. He was discomposed, and it seemed to Clive that if he could break the boy, he would not have to worry about the psychiatric case. The boy was the only witness to the actual crime.

"You say you saw the accused coming out of Deborah's house. Did you say anything to him?"

"No."

"I am recalling your testimony. You ran up to the house. You were facing him. Still you did not say anything to him in the doorway?"

"I was afraid to speak to him, in case he did something to me."

"Well, when did you say something?"

"I called when he disappeared into the dark."

"You had been standing there for some time?"

To'a shook his head. "As soon as he disappeared I called to Pila Naufahu to come."

"You were afraid to be near the man?"

"Yes."

"Yet you say you could tell he was the accused. What was he wearing?"

"He was wearing denim shorts and a dark colored shirt."

"Was he wearing glasses?"

"No."

This was pointless. Clive's questions only bolstered the witness. So he stood closer, facing To'a frankly.

"You said that you saw him before."

"Yes, often. I saw him with the other volunteer, 'Emili."

"How many times? Five?"

"More."

"Did you see him that week?"

"I don't remember."

"You don't remember. All right, but what day or week?"

"I don't remember the day or week."

"Did you see him in the daytime?"

To'a thought about what his shopkeeper father had told him. It does not matter what anyone says to you, a policeman, a lawyer, a judge, just say what you remember. You do not have to look at the person either. Listen to the question and answer only what you saw.

"Sometimes in day. Sometimes at night."

"In September?"

"I don't remember if I saw him in September."

"February?"

"I don't remember if I saw him in February."

"How about July?"

"I don't remember."

"You don't remember. And you don't remember. Did you ever speak to him?"

"No."

"But still you say—" Clive motioned at Dennis—"You saw the accused. From your house."

"I may have seen him from my house."

"You may have seen him. Your house is very far away from the house in the picture. It is thirty fathoms."

"No! No. It is not that far. It is not as far as the *'ovava* tree outside the court."

Clive returned to his table and looked at his notes. But he was well past his notes. Rude cries came from the market, also muted laughter, and the courtroom was still as a picture, frozen with feeling. The Tongans in stiff poses at the side of the room were praying for the boy to hold up. Even some *pālangis* had developed sympathy for him as he hung in through the cross-examination.

Clive stepped toward the prosecution table and took up the Seahorse almost daintily, holding it with thumb and forefinger by the steel knob.

"You were shown evidence at the lower court, young man. Do you remember that?"

"Yes."

"You were shown this knife. And you were doubtful about the knife, do you remember?"

To'a's wideset eyes welled with humiliation. "I was not doubtful."

"I am referring to page 6. 'Yes I saw a dagger beside the girl. There was blood. No. I did not see this knife before.' Do you remember those questions?"

To'a looked down.

"I was asked many times about the knife. By the Crown and you and the judge, then by you again."

"I was the lawyer for the defense, do you remember that?"

"Yes."

"And what you said is—let me remind you, you had not seen this knife before."

"I meant—I had not seen it before that night—"

Judge Hill held up a pink hand. Hill was new, but this was his court.

"Counsel, excuse me—I would like to understand what the witness is saying."

Clive dropped the knife on the Crown's table, and Hill looked down over his glasses at To'a, then at Kaveinga, the court interpreter. The judge had a simple and direct way.

"Young man, do you see the knife today that you saw on the night of October 14?"

"*Tamasi'i,*" Kaveinga began—boy, in Tongan—and Kaveinga was pulling for him with all his heart.

"Yes," To'a said.

"Where is it?"

"That is the knife."

"How do you know it is the knife?"

"Because I can remember it. There are notches on the blade. I saw the notches."

Hill made a note in his book with his fat fountain pen: "Identified the knife."

"Now, how can you be sure that you saw the accused that night?"

"I could see him clearly."

"Young man, you must explain to the court what you mean."

"*Tamasi'i—*" Kaveinga said. "*Ho'o 'uhinga?*"

For only the second time To'a looked at Dennis. Now the courthouse fell away—the polished rimu wood walls, the steel roof trusses with the silver metal stars, the crowd—into ether sunlight all spun and vanished. Gone were the stares of his countrymen. Gone too the bull in the black robe. To'a was again in darkness and the door was cracking open. Some small light now fell from the light bulb on the man with the beard, Dennis unfaded, Dennis muscular and tensed.

"Because I could see everything about him. His foot was holding open the door. I saw his forearms around her. He was holding Tepola by the chest. The muscles were standing out from his arm as he held her. He was using his fingers to grasp her."

"Thank you, young man," Hill said. "*Mālō au pito, tamasiʻi*," said Kaveinga, and Clive turned and retreated.

24

Tēvolo

He walked trembling and angry from the courthouse and that night his family feasted him and the news of his performance went out on the coconut wireless across the city, that Neti the greatest attorney in the Kingdom or outside it had tried to break the boy's word but To'a had broken Clive instead, and the Peace Corps with it. Barkow, the lawyer, cabled headquarters.

> DIFFICULT DAY FOR DENNIS, MAIN PROSECUTION WITNESSES ESTABLISH HIS PRESENCE AT SCENE.

And days 3 and 4 were no better. Several policemen testified. Faka'ilo described his interrogation of Dennis on Sunday after the murder.

"Is there another house located together with Mr. Emile Hons's house?"

"Yes," Dennis had said.

"Whose house is it?"

"I do not know."

"Do you know Miss Deb Gardner?"

"No."

Judge Hill wrote that down with his fat pen, and wrote LIES alongside it, also "Mens rea." Guilty mind.

ANOTHER DAMAGING DAY FOR DENNIS. EVIDENCE APPEARS VERY
STRONG. MY JUDGMENT, KEEPING IN MIND THAT I AM NOT
TONGAN IS THAT A STRONG CASE BUILT BY PROSECUTION. CLIVE
SEEMS TO AGREE. CLIVE GETTING SOMEWHAT TESTY IN COURT.

For a time, Clive performed his role, and made monkeys of the police witnesses. When a constable said he had seen a bloodstain on Deb's floor, Clive
objected, the officer was in no position to say it was blood, the stain had not been
tested. If a police officer referred to Deb's death, Clive objected that it was hearsay,
and Judge Hill agreed, and the officer had not a clue what he was allowed to say
and not to say.

But Clive lived in the real world, and the nitpicking now seemed to bore him.
He knew what the seven farmers in the jury box were thinking, Dennis had done
it, and his first line of defense having failed, he proceeded to put his energy into
the psychiatric case.

Dennis became enraged. In his view, the police were incompetent and the
case would have been tossed out of any respectable courtroom. He'd never really
believed in psychiatry, he did not believe that an insanity defense could work in
Tonga and did not think it necessary either. The case against him was a failure, he
thought that Clive should destroy the Tongan witnesses one by one, head on.

Then it was like every family he'd been a part of, Dennis questioned the interests of those who claimed to be on his side—Clive, Peace Corps/Washington,
even Peace Corps/Tonga. He said that everyone was deserting him, and he took a
swing at a policeman as he was led from the courtroom for lunch, and then said
that he wanted to fire Clive.

Laura had to calm him down at lunch, as did the psychiatrist, who had
arrived from Hawaii. The Peace Corps lawyer was dragged into it, too.

NASTY SCENE WAS AVOIDED. DENNIS STILL VERY UNPREDICTABLE
AND IS SEEMINGLY CONCERNED WITH ISSUES OTHER THAN TRIAL.

One issue was privacy. The legal process had breached Dennis's privacy again and again. For instance, he had seen a Tongan doctor on contract to the Peace Corps. Palu had prescribed Darvon to Dennis, Palu had then examined Dennis at the hospital on the night of the murder. Now Palu was called to testify. By what right did a Peace Corps doctor testify about Dennis's suicide attempt? Didn't the doctor-patient privilege bind the doctor to confidence?

The same went for the psychiatrist. Dennis had asked for the psychiatrist because he was in crisis. Now the psychiatrist was going to testify in an open court? Dennis had authorized no one to hear his statements but the doctor. PRIVEN REFUSES TO ALLOW DR TO TESTIFY ABOUT RESULTS EXAMINATION. His need to protect his privacy seemed greater than his desire to live; he did not see what his miserable sex life or his miserable mother had to do with a criminal issue.

The psychiatrist said that he would have to withdraw from the case. Otherwise he could be sued.

Laura's intervention was required to end the impasse. Home-cooking Laura was now the most trusted person in Dennis's life. She shuttled between Dennis's world and Clive's, and the psychiatrist's, too, and kept the team together. Dennis signed a release, the psychiatrist could testify.

Laura brought him his dinner that night, and a local weirdo with greasy hair came up outside the police station and declared, "Your brother will be freed."

There was something intemperate about Dennis's friendships. They were not made to survive everyday existence. Dennis could be the warmest friend in the world, but don't play games with him. See what happened when someone played games! So that meant that you must understand and respect his interests utterly. How many of the friends intended to be Dennis's friend when this trial was over? The love, the sway, the loyalty, the ferocity, they were extraterritorial, specific to this reality, the Peace Corps experience in Tonga. Back home safe in America, how many of these people would want to continue an alliance with a crazy man?

The other close friends, the Bird Lady and the Pauls, were not in the courthouse. They were waiting to testify. They and the oldest friend Emile sat on the grass or on the knees of the banyan tree across the street from the courthouse.

Emile wore his pink tie and those scratchy wool bellbottoms with the dull red stripe, and now and then he and the Bird Lady walked up to the courthouse windows to try to see what was happening. Till the judge yelled at them through the window, and a policeman asked them to step out of earshot.

Then it was Emile's turn to take the stand. He testified for two hours, and felt torn between the friend in the dock and the girl whose body he'd taken home.

He identified the broken glasses as Dennis's, and the Seahorse knife, as well. But the Crown had misplaced the knife sheath also found in the house. So it asked for a recess and sent out for the sheath, and a police jeep came roaring down One-Way Road to deliver it, a policewoman leapt from the vehicle and sprinted across the grass to the courthouse with the sheath, which Emile then identified.

Under cross-examination, Clive turned Emile to Dennis's state of mind. Emile said that Dennis had been depressed the week of the murder, and then Clive made a lawyer's mistake, he asked Emile a question to which he did not know the answer.

"Did the accused ever show hostility to the deceased?"

Something came to Emile's mind, he almost lost his breath.

"I do remember once, a few months back. Dennis and Debbie and I were playing cards in my *fale*. We were all sitting on the floor. Dennis never lost at cards, then Deb finally won a round and stuck out her tongue at Dennis. It was completely lighthearted, but Dennis lost his temper, and threw his leg out at her, kicked her so hard he knocked her over."

His foot had shot out like lightning, and Deb and Emile had been shocked.

"Because other than that Dennis was always gentle with her."

Emile left the stand without looking at his old friend again, and giving Zenker his pink tie, rode his bicycle back to Ngele'ia for almost the last time.

Clive was to open his defense on Friday, December 14, and on Thursday night he went to his cousin's house and sat with three of his uncles to strategize.

"I must show that Dennis is *fakasesele*," he said, using the Tongan word for crazy. "If I can show that, the rope is over there—"

Clive pointed at the floor in the corner of the room.

"But if I cannot prove *fakasesele*, then the rope is here."

Clive held his hand around his neck.

He was back on his heels now, but that was all right, Clive had reservoirs of swagger and cunning that he had never before exhausted and would not exhaust that week. From the time of the preliminary hearing, he had known that the eyewitnesses were strong, and yes, he had hoped to damage them, and he had damaged the woman. But the boy? He could do nothing with a boy like that but sit down. He should have sat down much sooner.

"Someone get me the Bible. In Tongan."

Clive made his uncles take turns reading from the Gospels, and covered his eyes with his hand to listen, and now and then directed them to different passages, and the next day he carried that same black leather-bound Bible out to the middle of the courtroom.

"The defense will be in two parts.

"First, the accused denies that he is responsible for Deborah's death. No one has testified that he saw the accused killing Deborah. The prosecution has offered only a circumstantial case.

"But even if you do not accept that defense, and you find that the accused caused her death, you must find that he suffered from an illness of the brain, and did not know what he was doing. A doctor of the mind has come to Tonga all the way from Hawaii to testify. A doctor who has examined the accused."

Clive opened the Bible, and faced the seven farmers and no one else.

"Luke. Chapter 23, verse 34. 'And Jesus said, forgive them Lord, for they know not what they do.' Members of the jury, you must remember what Jesus said on the cross when you listen to the defense. Forgive the man who knows not what he does."

Clive spoke in Tongan, and used a simple language that the jurors might hear in church. There was a law in Tonga's three black books of law, if a man was mentally sick and did not know that what he was doing was wrong, then he could not be convicted of a crime. Kupu 17: Section 17. In order to convict, the law requires heart and mind. Not just one, both. *Loto* and *'ilo*. Intent to do it, knowledge of what he was doing. If the mind was sick, if that knowledge was not there, the jury could not convict.

"Of course you see the accused in the courtroom. He will not give testimony. There is no point in calling Dennis to the stand. For he has no memory at all of the incident."

Then the defense called its first witness—Sinoveti Motuliki, Clive announced her name.

Through the door nearest the jury now came a slight, sinuous woman, so small that those in the back could not see her over the heads of others. She smiled at Dennis, and Dennis smiled back at his Tongan neighbor, and Sinoveti sat comfortably in the witness stand. With her long coiling dark hair and glowing eyes, S'no was a kind of performer, and she wrapped one end of her hair around her fingers and was not nervous.

"How is it that you know the accused?"

"For two years Dennis has been my neighbor. He was a good neighbor. Sometimes we went to the movies together."

"Can you point him out to me now?"

"Yes."

Typically in Peace Corps service, a host country family kept an eye on a volunteer. Dennis's former *fale* was on the Motulikis' land, and Dennis had often helped the Motulikis, helped the three boys in math, and helped Sinoveti, a primary school teacher, prepare herself for the examination she would have to pass to teach at a higher level.

"Did there come a time when you stopped associating with Dennis?"

"Yes."

"Why was that, please tell the court."

"He drew a skull on his door—"

S'no was carried along by the story.

"I said, 'Dennis, why have you painted that on your door?' He explained that the devil came in at night and sat on his stomach and seized him by the neck. It was a female devil in appearance but when it sat on his stomach it appeared like a man. He put the skull so that the devil would stay outside."

"And then? Tell the court."

Dennis put out plates of food, papaya and taro, because if the devil ate the food, maybe it wouldn't bother him, and he cut his door in two, with locks on both segments so that he could fasten the bottom and reach out from above and push the devil out and it couldn't crawl back in.

"Dennis, this is a devil—*tēvolo*—you must get rid of it with Tongan medicine," S'no said.

"I don't want any Tongan medicine," Dennis said.

It was July. S'no came home from school to find Dennis in the yard gazing into the trees.

"Why does that tree have flowers while that tree has none?"

"That is the *mei* tree," S'no said. "It has already bloomed and made its fruit, you can see the breadfruits now, while this is a banana tree. Now is its time."

Then S'no went to the bush to get the Tongan medicines, and made a bundle of the leaves, and held them up to Dennis's face. Dennis cried out and ran into the house, which confirmed S'no's belief that Dennis was *'avanga*. *'Avanga* was a Tongan concept of possession. A *tēvolo* was inside him, he was in danger of hurting himself.

S'no had seen the same *tēvolo* herself. The only way to cure Dennis was for someone to hold him down and squeeze juice from a paste of *uhi, nonu* and *lautolu* leaves into his eyes and ears and mouth. S'no could not do that herself. Still, she went to Dennis's house and called on the *tēvolo* to leave him, and threw the plants through the door.

Later she saw he had thrown the medicines into the yard.

"After that everything was changed. It was as if he was not really my neighbor anymore."

"I have no further questions."

Tevita had been many years in Auckland and London, but he could see the threat. Clive was infiltrating a Tongan folk understanding of mental illness into the *pālangi* case to soften up the jury so that they'd accept the Western mumbo jumbo that was sure to follow, and Tevita needed to expose that at once. He proceeded with a kind of sharp patience, pricking S'no point by point.

"I would like to call your attention to the picture on the door. Why is it that the accused drew the picture?"

"So the *tēvolo* would stay in the picture and not come inside."

"You talked to him about the picture—how many times?"

"Twice about the picture."

"And you believed he was telling you the truth, why is that?"

"Because he was serious and sincere."

"Do you go to church?"

"Yes."

"And which church is that?"

"The Wesleyan church."

"But you believe in *tēvolos*?"

"Yes."

"Now you said that you had also seen this *tēvolo*."

"It came to my house."

"The same *tēvolo* or a different *tēvolo*?"

"The same *tēvolo*."

"Please describe to the court how the *tēvolo* appeared to you."

"It came as a woman. It reached out to me, and I hit at it. There were sparks flying."

A devil with sparks flying, that was very good indeed, and Tevita nodded, with a faintly arch expression.

"Did you tell anyone about this?"

"I told my children and the neighbors."

"Did the *tēvolo* come to your house before you had talked to Dennis about it?"

"No, after."

"Did you draw a skull on your door?"

His irony went right by S'no.

"No. I brought in some medicines. Because I knew that would keep the spirits away."

"Because this *tēvolo* had come to you before?"

"Not this *tēvolo*."

"But other *tēvolos*. How often?"

"Quite often."

"And when was the last time that you were visited by the *tēvolo*?"

Finally it occurred to S'no that Tevita was having his way with her.

"I am not sure when," she said.

"This year."

"Yes."

"And did you tell anyone about those *tēvolos*?"

"I told my husband. I told Dennis."

"When did you tell Dennis that you were visited by a *tēvolo*?"

"When he talked to me about the *tēvolo*."

"You did not tell him before that?"

"He had not asked."

"But now—he asked?"

"Yes, and I told him it was quite common for *tēvolos* to visit Tongans, I told him about prior visits. Because I did not know that the *tēvolo* could come to *pālangis* as it came to Tongans."

It was a fine cross-examination, surgical and neat, using humor and S'no's vanity to undermine her. The testimony could not be trusted, she had either imagined the whole thing or the clever defendant had picked up the *tēvolo* talk from her and parroted it back. Michael Barkow cabled Washington with grim news.

WITNESS ADMITTED UNDER CROSS EXAMINATION THAT SHE HAS HERSELF BEEN VISITED BY DEVILS.

But this was the problem with the cross-examination, Tevita was playing to the educated. That was his weakness. Having been chosen in life, by this powerful person or that one, he lacked the toughness and vernacular style that made a great courtroom performer. As Clive had, he had worked in the slaughterhouses in Auckland to pay his way through school, but Clive had liked the work, shouldering the carcasses and getting bloody, while Tevita had preferred the freezing room, because it was clean, no guts around, wiping down the lambs just before they were frozen.

The jury were men of the soil. 'Aminiasi, Feki, Hesekaia, Kepueli, Leimoni, Loloma, Sione—they were farmers, though one was also a steward in the Wesleyan church, Clive's secret agents had informed him of that. But only two or three had even made it to high school before their families had directed them to get to work in the bush. They could read. Nearly everyone in Tonga could read. But none of them spoke more than fragments of English, and to them S'no was not crazy. "Tēvolos are common knowledge in Kolomotu'a," S'no said truly, referring to the old city of Nuku'alofa. Shamans still made house calls, and made pastes of leaves, and dripped drops into eyes, nostrils, mouths, to get a tēvolo out.

As for Christianity, there were times when Jesus had exorcised demons, and so it was common for Tongan ministers to fold tēvolo talk right into their sermons. A person could as soon dislodge a Tongan's belief in tēvolo as his belief in moonlight. That went for professionals, too, like Kaveinga, the interpreter. Kaveinga was a true intellectual. He had read history at the University of the South Pacific in Fiji, and before the trial had bought two books to give him a grasp of the obscure Western terms that were likely to be introduced. A medical encyclopedia, a medical dictionary— Kaveinga had studied all the portions relating to mental illness. But he had been many times himself to folk healers, he believed in tēvolos and 'avanga too.

Some English words overlap with 'avanga—possession, infatuation, dissociation, even schizophrenia—but they don't define it. 'Avanga means being carried away by a spirit. Maybe the 'avanga person has become obsessed with a dead person, maybe he has gone too close to a graveyard, brushed up against a spirit with unfinished business. But something from the other side has taken that person's mind.

'Avanga causes him to flee society. Most commonly he runs to the sea or to the bush. Four or five men are required to hold the 'avanga person down, and later he might remember nothing of what he had done or said while possessed.

So Sinoveti's story gave a Tongan underpinning to the story Clive had already begun telling. Dennis had become withdrawn and depressed, he had rejected society. That night the pālangis had gone to look for him on the beach and in the bush.

When he came to the jail he was afraid of the darkness inside. He remembered nothing of that night—or so Clive said.

Could a *pālangi* suffer *'avanga*? "I did not know *tēvolos* could come to *pālangis* as to Tongans," said S'no. Well, now she knew. Dennis had been here two years, he had lived near the bush in Longolongo, and a *tēvolo* had come to him.

Tevita's cross-examination went on too long, or it became too analytical, and put the prosecution at odds with folk belief.

"You said that Dennis's *tēvolo* visited you just once. But not again?"

"No."

"Before?"

"This *tēvolo* had not come before."

Tevita said, "But other *tēvolos*?"

"Yes."

"And what did you do to chase away your *tēvolo*?"

Naughty laughter rippled the courtroom, it was almost as if someone had sat down on a whoopie cushion. Tongans in the gallery laughed, the farmers' eyes sparkled in the jury box, Kaveinga smiled at his desk even as he translated the Tongan into English and, peeking at Clive, exchanged a grin.

You won't find "your devil" in the big Tongan dictionary written by the missionary because it's slang for the genitals. As for the expression "chasing your devil," it also has a distinct meaning. Earnestly and precisely, Tevita had tumbled into that staple of Tongan humor, the masturbation joke.

25
___

The Verdict

A day after the murder, Peace Corps/Washington had called a psychiatrist in Hawaii who had once done volunteer screenings, and Furukawa said there was only one man for the job, a forensics man he knew, and Furukawa called Stojanovich, who said, "Tonga, where is Tonga?" "Polynesia—" And Stojanovich hummed in his meditative way.

He loved Polynesia, Polynesia had liberated Stojanovich from the old world, and on Sunday after the call he had headed south from Honolulu and though he would have liked to bring his surfboard he thought it inappropriate. Word gets around in no time in a Polynesian community. It would not advance his cause if the Tongans saw the sun-loving hedonist in him, the surfer with the muscular body.

He'd been stunned on arriving. This is a capital city and ancient monarchy? It's just a car or two on the road, and how can this be a Kingdom and how can they have a king when everything seems to be so small, and in this deflated expectation he was soon friends with Mary, another cosmopolite who had led a hard life before landing here.

No pad, no book, no tape recorder, Stojanovich brought nothing into the cell with him. He had everything he needed in his little finger. Everything was in his life experience. He did not need to consult *The Interpretation of Dreams* or the *Diagnostic and Statistical Manual* to know in an instant what was crumpled before him.

The setting was exotic, the patient was not. He was run-of-the-mill, bedraggled and pushed down inside himself, withdrawn and taciturn. "My name is Kosta Stojanovich and I am a psychiatrist." But the man did not want to take his hand, did not want to be touched or physically involved with someone, and when they began to talk, he looked to the left and to the right, rather than to Stojanovich's face. Simple schizophrenia.

The conversation was also schizophrenic. Dennis meandered left and right and right and left about various things, from Mary, whom he hated deeply, to what awaited him, but never staying on a subject, keeping little connection to reality.

Outside, Stojanovich could hear the big Polynesian policemen laughing and mellow, inside it took him two or three hours to get past Dennis's defenses. Dennis did not want to talk about his family. So Stojanovich skirted the family talk and the girl as well, until Dennis volunteered things. Not that he trusted Stojanovich. Trust means nothing to a schizophrenic. But he did become somewhat comfortable.

His relationship to his family was very strained. He had no close ties with any member of the family. He did not want to be more specific. His mother had done things to upset him, but he did not want to elaborate. This was unusual, most schizophrenics would say, well she was OK, but Dennis became agitated and said, "Don't push me into it."

"I will see you in two days."

And it was a relief to Stojanovich to be back in the fresh air. He did not like confinement, hadn't since he was a boy.

He had grown up in an elite section of Belgrade. His father was a general, the king had provided an opulent house. Then the war came. First the Germans put his father in a concentration camp, after that Tito incarcerated him because of his associations with earlier regimes. Then the so-called Allies bombed Belgrade, and until you have lived in a city that is being bombed, do not prate about war and justice—the boy saw people moaning and helpless in the streets, their guts trailing out of them.

Then his father disappeared. Kosta was the last to see him. A boy of 9 visiting the hospital, he saw his father one day and the next day he was gone.

His well-connected sister brought the family to the States, where Kosta earned one degree after another. West and farther west: he wanted away from these organized modern places. Hawaii had a Polynesian ideal of community and the pleasure principle, he went to Hawaii, where he worked for the state at first, testifying, The accused was sane. But it did not suit Stojanovich to work for the state.

A state served a group, and different laws served different groups, laws in no way approximated justice. You could not write down justice on a piece of paper. Justice was personal and emotional, an ideal of fairness. Law was about power. He'd seen what states did on the streets of Belgrade. So he went to work for defense lawyers, and out of ten times that he entered a cell, six or seven times he said, The man is crazy.

The second day with Dennis went better, October 21. One week had elapsed, Stojanovich had spent hours with the Pauls, and Dennis seemed to grant the existence of psychiatry.

These things the police say you left in Deb's house, the bottle of cyanide, the syringe—did you have an unconscious desire to leave them? You planned to kill yourself with an injection of cyanide, then you left it there. "Maybe you actually didn't want to kill yourself?"

At night Stojanovich went out to eat with Mary. That was a pleasure. Clive had impressed on him the need for secrecy. "No one is sure what you are doing here and let us keep it that way." He wore a bright Aloha shirt and seemed to be on vacation. But Tongans noticed him and were sure he was CIA.

He had dropped the disguise for the trial. He wore a light jacket, a white shirt, a tie, sat bent slightly forward, testifying in a soft sweetly accented voice. He was five-foot-eight, and if you didn't look into his eyes you might think him weak, but his eyes were piercing blue and level, and he looked something like the actor Anthony Hopkins.

"And your conclusions?" Clive said.

"My conclusions were that the accused was suffering from a serious mental disease called schizophrenia. For most of his life from all the evidence I gathered he has suffered from latent schizophrenia.

"This condition under stress would deteriorate very much and very fast and he would go into a state of depression, and this resulted in a breakdown, a state of acute paranoid schizophrenia."

"What in your opinion were the stresses?"

"Mary George has said that he had a strong concern about his future. Would he be staying on in Tonga or Fiji? He did not know. Paul Boucher, whom he considered his closest friend, has described a withdrawal by the defendant. This would be a symptom. The defendant has told me that he had been depressed for about two months prior to October 14, 1976."

"And the breakdown? What were the signs of that?"

"Delusions. A delusion is a false belief, a fixed thought that cannot be corrected by reasoning. A series of other misrepresentations follow, resulting in an acute breakdown, where he does not know that what he is doing is wrong."

Whenever Stojanovich said "schizophrenia," "delusion," "megalomania," or "paranoia," Kaveinga was compelled to repeat the English word in his Tongan translation. He tried to help the farmers out with Tongan definitions, but there were no Tongan definitions, even with the medical dictionaries. The closest he could get to paranoid schizophrenia, was *anga-ua*, which means to have a double personality, quixotic, lacking in straightforwardness. *Anga-ua* had never been used to excuse a murder, so there were other terms that Kaveinga resorted to as well. *Fakasesele* meant eccentric or crazy. *'Atamai vaivai* meant weak mind, a crazy person. *'Atamai maumau* was broken mind. *'Avanga* suggested possession. Any and all would have to serve.

"Doctor, did you observe these delusions in the accused?"

"I did. He appeared to be in a dream world. I observed an absence of feelings, or the feelings he did display were inappropriate. It was as though his feelings were out of reach. At important questions he would laugh as though they were unimportant, and at unimportant questions he became sullen, serious, withdrawn.

"You need to be from the same cultural background to assess feelings. Most Europeans will not kill animals for their food. They will have someone else do that. But he said that he had killed animals, and when he did it, it was as if someone else were doing it."

Two hours had gone by, and not a mention of Deborah Gardner.

"Tell us about Dennis's feelings about women."

"The patient was afraid of women. He found it difficult to approach women sexually. I believe that he has had only three or four sexual experiences in life, and that these were devoid of feelings."

That came from his mother, Stojanovich said. There was no love, he was afraid of communicating with her. "When he had opened up to her, she had humiliated him and he feared that other women would do the same."

"And his feelings toward the victim in the case?"

Judge Hill tilted forward.

"It is late. We will resume again tomorrow."

That night the Bird Lady saw her neighbor, and 'Uini had an encouraging insight. She had heard about S'no's testimony about *tēvolos* and 'Uini concluded that Tenisi would get off. Tongan juries didn't like the police, didn't want to hang anyone anyway.

'Uini had pedalled around town on the Bird Lady's bicycle that day, getting official signatures for the clearance form the Bird Lady would have to present at the airport in order to leave the country. She was going. She had thought of doing a third year, but the case had taken the will out of her. She had mailed thirteen packages home and went around town in a fragile mood, regarding each perfect piece of the terrain and wondering how she could ever leave it.

The coconuts like tightly wrapped bouquets, black against the swollen orange moon. The massive mango trees like little hills, their dense leaves hiding the birds. The puritan branches of the kapok tree, going out at right angles. The rain trees creating broad rooms, their small bipinnate leaves barely filtering the light. While the sea was so still it seemed almost as if it weren't there, a pale shiny color that passed seamlessly into sky, far too pale for anything as strong as the sea.

Then she reflected that she had left the red sumacs of Michigan, the flowers in Anderson's woods, the twisted spruces on the Oregon cliffs, and in that way consoled herself, that memory could hold all these things, however imperfectly.

The Bird Lady was able to visit Dennis for five minutes after Stojanovich's first day. A policeman looked into a lawbook and, deciding that a witness was allowed to visit a prisoner, let her into cell 1. He was supposed to stay and monitor the two but left them there together as if to facilitate a conjugal visit. They were both a bit embarrassed, standing in the dark cell.

Though Dennis still had his sense of humor. He was glad that Emile had managed to get into the trial record the fact that he'd always won at poker. . . .

The Bird Lady was glad that Stojanovich's testimony about the unconscious seemed to interest Dennis. So maybe there was hope for Dennis, maybe his life was not ruined. The Bird Lady had some understanding of Dennis's gift, he could hollow out another reality right inside consensual reality and carry people into that warm cave on an underground river, and they were enchanted, at least for a time, and he was so intelligent and stubborn and oversensitive that it would take someone of great determination to challenge that place's existence. That someone was not her, only a psychiatrist back in the United States would have that power. Unless Dennis bridled and refused to cooperate, which of course Dennis could do.

He might be on her flight. She'd made a reservation for the Fiji flight on Thursday afternoon, December 23, and even if Dennis was convicted, the Tongan secretary of government had suggested that the Parliament might pass legislation to allow him to serve his sentence elsewhere.

Barkow, the Peace Corps lawyer, had heard an even more encouraging report.

HAVE HEARD RUMORS FROM VARIOUS CIRCLES THAT PRIVY
COUNCIL WANTS TO GET DENNIS OUT OF TONGA IRRESPECTIVE OF
VERDICT. MY UNDERSTANDING IS THAT COUNCIL PREPARED TO SIT
IN EXTRAORDINARY SESSION AT CONCLUSION OF TRIAL.

Just about everyone in Peace Corps was now hopeful that this distraction would disappear, and when a group of volunteers ran into Rick Nathanson at John's Takeaway in the King's Road and he said that he was telexing reports on the case to Australia, the others got angry, and murmured, "Asshole" and "Worm," and pointed out that what he was doing could only hurt Peace Corps. "No one appreciates the press till it's not there," Rick said, then quoted some Thomas Jefferson he'd studied in college: "Given the choice between a government with no newspapers and newspapers with no government, I would choose the latter."

◆   ◆   ◆

On Tuesday morning Tongans crowded the courtroom for the return of the star witness whose voice was almost a purr. He was the most educated man most Tongans had ever seen, and his ideas were dazzling, he was describing a way of looking at the mind that they had never heard of. And though the Tongans were baffled by Dennis, baffled that he had said nothing at all through a week of trial, only now and then rolling his head in exasperation or baring that tiny smile through the heavy beard, and though they were still hopeful that he would make some statement of his actions, as custom compelled a Tongan to do, they understood that the eloquent psychiatrist might be all they would get.

Stojanovich sat down, and Judge Hill reminded him of his oath of the day before, and Clive picked up where he'd left off, the question about Deb.

"I would say that it was a very unusual and strange relationship. The patient was not in love with her. I mean that there was no love as reasonable men experience it. But he did not have hateful feelings toward her either. The patient was disturbed by the victim's behavior, and he strived for a spiritual union with her. That sort of delusion is always associated with schizophrenia."

The psychiatrist had heard a story from volunteers involving Deb's body. A group of volunteers was out on Yellow Pier. It was dusk but hot, the water glinted, and Deb Gardner the tomboy pulled her sundress over her head and jumped into the sea. Anyone could see her flesh, and this had shaken Dennis. He yelled something at her, it was *tapu, tapu,* culturally insensitive and wrong. Then he had walked away, burning up and burning up again over this sexual woman's behavior in the presence of men.

Now the psychiatrist offered a formulation of these feelings that Dennis's friends had not heard before.

"Can you be precise about the character of these delusions?" Clive said.

"I can. For a few months prior to her death he thought that she was possessed by the devil, and he became convinced that he could save her spiritually, that he would be her savior. He thought he was the only one who could expel the devil from her.

"So there were three delusions. Number 1: He thought that Christ fought devils and saved the people. 2: He thought that he would fight the devils in her even if he had to die himself. 3: He thought, I am therefore her Jesus Christ and savior."

The stillness in the courtroom broke. A murmur began, a cry or two, a shout. Clive stopped and let it all die down.

"Would it be consistent with these delusions for the defendant to have attacked the deceased?"

"Yes. He would have thought that he was saving her. That he was having a spiritual union with her."

"Do you know whether the defendant killed the deceased?"

Stojanovich shook his head.

"I do not. He does not remember. He has many gaps in his memory. This is typical of a breakdown. Some of these gaps are over the last several months, but the longest gap is on the night of October 14. Everything is a fog. He remembered going toward the house of the victim. From that point on everything is in a cloud. Until he found himself at Paul Boucher's house, and they cycled over to the police station, where he is afraid of the darkness, he does not know what he is there for—"

"I have no further questions."

The judge picked up his gavel.

"We will take a recess before the Crown begins its cross-examination."

Clive had set out two months before to sidetrack the prosecution. Now that work was accomplished. A trial is a story, Clive had told a compelling one. Dr. Stojanovich's devil talk mixed with the *tēvolo* talk of Sinoveti before him, and the talk in the volunteers' testimony about Deb's freedom and boyfriends. Demons mingled and danced, from psychiatry to Christianity to counterculture.

Tevita and Christine had controlled the facts of the case, leaving Clive to grasp at straws, but they had not seen the dark horizon of the trial, they had not come up with a story about Dennis's motivation. A schoolboy had been their most important witness, and they had looked on the psychological dimensions of that night in everyday terms. Dennis was a chemistry teacher who had taught in school that very day. His students had noticed nothing in class. He was seeking a third year as a volunteer. He was coherent.

Stojanovich was both more outlandish and more effective than the prosecutors had ever imagined, and now the Crown was demoralized. Many times before he had performed on witness stands, he had a soothing nonargumentative tone. Oh yes, they had been overmatched, the American government was spending, by one estimate, $100,000 on Dennis's defense. Tevita's salary for the entire year was just half of what Clive was to receive for his services over a couple of weeks.

All the same, Tevita and Christine could react. They could see the weaknesses in the psychiatrist's amazing story, and they felt that it was canned. No one had heard a thing about any of these delusions. As Christine whispered to Tevita, the legal issue at the case's heart was not whether Dennis was mentally diseased, a paranoid schizophrenic, call him what you will. It was whether the disease had caused a defect in reasoning so that he did not know whether what he was doing was right or wrong. Well, he knew. He had been working and playing with others, teaching school. He had turned himself into the police, divided his possessions. Being Jesus Christ—no one had heard a word about that.

Tevita was aggressive.

"You spent all of eight hours with the patient, and most of that time he was unresponsive. Then suddenly he tells you these delusions."

Stojanovich did not seem to have a defensive bone in his body.

"Yes. I do wish that I had gotten more details from the defendant. He was very resistant to talking. He had just had an acute breakdown. To see his mind was to look through a fog."

"If the accused was so deluded, how come none of his friends had known about it?"

"That is very typical of the latent paranoid schizophrenic. His behavior seems normal to his friends. He does not show this side of himself to his friends."

But Tevita went down the list of *pālangis* who had noticed nothing really amiss, the Bouchers, Vickie, Emile, Frank.

The psychiatrist adopted a patient tone.

"You must understand the mind of the accused. He took great pains to resemble a normal person as much as possible. He is like a volcano. People can play on the sides of a dormant volcano and have little idea of what is going on inside."

"But the two doctors who treated him that night noticed nothing odd. Lutui said he was fully conscious, answering questions clearly."

Stojanovich nodded.

"The gaps of memory do not last for a long time. For several minutes or in rare instances for half an hour. Then he gathers himself and returns to a latent state."

"But a person who has killed someone would be fearful of entering a police station—that is completely natural."

"Counsel, a delusion is like a dream. In a dream we can hear, see and touch . . ."

Then Hill broke in on the questioning. The old army major was not much on psychology.

"Do you have any direct evidence that the accused killed the victim?"

"I do not. At a certain point he may as well have been an automaton."

"But where exactly does that begin?"

"He was bicycling toward her house. That is the last thing he remembers."

Or maybe not, of course.

That was the most stupendous leap in the testimony. If Dennis had forgotten everything, how was it that he remembered "the struggle" at Deb's *fale?* Whether or not Peace Corps wanted to know about it, it had been informed on this point several times. Clive had personally told the associate counsel about "the struggle," and Magid had made careful note of it in his legal pad, then typed that point up when he got home. Clive had told Dennis's parents about the struggle, too, and copied that letter to Washington.

If everything was in a fog, then how come Dennis had told Emile on Thursday that he was planning to go to Deb's house at nine-thirty that night—another point recorded in the parallel investigation back in Washington? If he had forgotten everything, how was it that he had spoken to friends of the messiness of that night, the messiness that was so unlike him, the lack of control? Apparently that had seemed to Dennis the insane part, going nuts at Deb's house, going over there and not saying a word, not trying to have sex with her, just losing control.

And if he had been in a fog till he got to the Bouchers, then how come he remembered taking pills at his house? The psychiatrist had discussed that with him.

Did his leaving the cyanide on Deb's floor mean that he did not want to kill himself?

There had been no fog, Tevita and Christine sensed that. And though they hammered helplessly at Stojanovich, with Christine asking him why he had not used truth serum on the patient, they were at a loss. They did not have the information, Peace Corps did, Peace Corps officials and volunteers, and Peace Corps was merely monitoring the situation.

At the end of the first day of cross-examination, Barkow sent a confident cable back to Washington.

HARD TO GAUGE EFFECT ON TONGAN JURY ALTHOUGH THE JUDGE
IS VERY INTERESTED PERIOD HE SEEMS TO FEEL THIS IS THE CRUX
OF THE CASE . . .

The next morning, Wednesday, December 22, the last best hope for the prosecution left the country. The Polynesian Air flight for Samoa took off with Vickie and Emile aboard.

Vickie Redpath was terminating after one year. She had done her job teaching teachers about grammar and Shakespeare and the Romantic poets, but she had never developed any real love for this place. Her home was the Blue Mountains, and this was some interruption she called "Togo," a typographical error in the dictionary of the world.

It was said that Vickie couldn't wait to go, that after October 14, she felt unsafe there. The case had dropped a bomb on Tonga 16. Dave and Kay Johnson had lately gone back to California. Dave said this place was too weird, and Dave should know, the Johnsons had been right in the middle of it. They'd seen Dennis the night before the murder, they'd had Dennis and Emile over to dinner then gone out to see *The Man Who Would Be King*. Dennis had said a couple of strange things that night. "I'm going to make Deb love me, whatever it takes." Well, what was he going to do? Francis Lundy had heard threats, Emile had heard threats, too.

Vickie had been the prosecution's strongest witness. Other *pālangis* may have escaped the Crown, but Vickie had opened a window on Dennis's motivation. Vickie was serious and literal. She had studied her Peace Corps oath for a long time before signing it, and she had sworn her oath on the Bible with the same grim seriousness before describing a real human backdrop to the murder. There was a triangle, Dennis was tormented. He did not know where his friends stood, he felt Emile had taken advantage of Deb after the Dateline dance. Vickie's testimony undercut Stojanovich's, and suggested that Dennis was possessed by feelings of jealousy, betrayal, and rage, feelings that did not legally excuse murder.

But when Emile had come to the stand, Tevita had asked about his relationship to Deb, and he had said that he and Deb were brother and sister, it was not a sexual relationship.

"What happened after the Dateline dance?"

"I looked after her that night," Emile said. "I told Dennis I looked after her that night."

Emile and Deb were free spirits, they were neighbors and friends, and they were lovers, too, and Dennis who hated sexual women had learned something the last week of Deb's life that set him off. Involving Emile, no less! A close friend. He had loved Emile in his way, and Emile had loved him, too.

Dennis had asked Emile the same question Tevita later asked him under oath, and as Emile had deflected Dennis in the street, or tried to anyway, he had deflected Tevita on the stand. A brother-sister relationship. That was a lie. Emile felt protective of Deb. He felt there must be some zone of personal privacy, that whatever had gone on between himself and Deb was no one else's business. There had been damaged feelings all around.

So the triangle remained submerged, and Emile was going. His two years were up. He wanted to make it back to San Bruno by Christmas Eve to surprise his parents. Some people had said that he should wait for the verdict, but why did he want to do that?

The last weekend had gone by in a blur. He took friends out to the Dateline for drinks, he had a sort of going-away party. His friend Tomasi had come back from the farthest island, and they tied one on. Tomasi was staying on for a third year, or forever, who knew. Then Tomasi and Emile walked down the Vuna Road along the sea to Joe's Place, and Emile brought a girl along from Yellow Pier, and they had more drinks, and some time that night the girl began passionately kissing Tomasi, and Tomasi looked over at Emile for guidance, but Emile, who was hip and urbane, simply waved his hand.

"When she gets drunk she likes to kiss."

He did not say goodbye to Dennis. His last meeting with Dennis in the jail had been too upsetting. It seemed that he had taken leave of this dear friend a long time ago.

Vickie said goodbye. She went to see Dennis a day or two before she left, to wish him luck. She brought Francis Lundy along, though Francis was against it. "If you do this, he's going to think you're his friend, you're giving the wrong impression."

But in the end Francis went along, and said, "How you doin'?" and "Oh really," and other nice things to Dennis in the jail. Then the Bird Lady came in and Dennis chattered away in his intense and persecuted way, as if some absurd series of foul-ups by Peace Corps and the Tongan government had landed him in this bizarre predicament, and a police officer said something about Tongan juries generally staying out for only five or ten minutes, the Bird Lady said, "God bless you" and they all left, and Francis was not sure what he, Francis, was doing there, or any of the other volunteers. It just did not seem real, and he wondered how sincere any of these people were in their friendship.

Now the Polynesian Air propellers beat the thick air of the rainy season, Vickie and Emile looked down for a last time at this period in the map of the Pacific, and far below the psychiatrist told an astonishing story about Christ and the devil.

The cross-examination went on all day Wednesday, Stojanovich had never been on the stand so long. The last defense witnesses sat waiting out in the grass, but Stojanovich was taking so long it didn't appear that the defense was going to call either one of them.

Paul Zenker wore Emile's pink tie and a suit jacket from the other Paul, and walked around prepping himself on points he was determined to make as Dennis's character witness. The Bird Lady got a good tan and brought her bags to the Peace Corps office, and gave her bicycle to Tomasi.

Then Judge Hill wearied of the mumbo jumbo and fell asleep. For a little while it was not clear whether the judge was asleep—his head was just tilted forward over the bench, his eyes shut—then the head gave another little lurch and the glasses slapped onto the bench and woke him up, and the judge grew red-faced with embarrassment and gaveled a recess.

When they came back into the court, Tevita was desperate, he had to do whatever he could to break this cool little shrink.

"I have in my hand a Bible, have you read this book?"

"From cover to cover."

"And in this book, the Gospels are about the life of Jesus. You have read the Gospels?"

"Of course."

"Doctor, I ask you. In the Gospels, in the life of Jesus as told by Matthew, Mark, Luke, and John, have you read anywhere that Jesus is described using a dagger to remove devils from anybody?"

"I have not."

"No, you have not. Because the Bible does not say such a thing, doctor. Nowhere in the life of Jesus does he remove the devils with a dagger, is that correct?"

"Yes."

And Tevita turned away, having made his point that the Jesus talk was a fantasy, concocted by a cold-blooded killer who had only had the Bible to read in jail—or maybe by a cold-blooded psychologist who had been a witness many times.

Till Stojanovich said, in that sweet purr, "Counsel, may I ask you a question?"

Tevita looked to the judge. Hill looked to the witness.

"Go ahead," said the judge.

"I must ask you the same question. Have you read the Bible?"

"Yes."

"And you are familiar with the life of Jesus as described in the Gospels."

Tevita said nothing now. The courtroom tensed, and even Clive looked on with interest, never having seen a witness examine a prosecutor.

"And do you know where it says anywhere in the Gospels that Jesus is suffering from paranoid schizophrenia?"

Tevita sat down, still the doctor held the court, answering his own question.

"No, and that is my point, the accused is not living in our world."

⋆   ⋆   ⋆

By the time Stojanovich was excused, it was Thursday, December 23, the longest trial in Tongan history, or Tongan memory, which is the same thing, and in a final flippant gesture, Clive called not Zenker or the Bird Lady but a German trader to the stand, to say that he did not sell flip-flops like the one in evidence, with the thong a different color from the sole. As if to say the police had doctored the evidence from Deb and Dennis's doorways.

And with that the defense closed, the cases were summarized, and Henry H. Hill at last took over his courtroom.

"Members of the jury, the time has now come for me to carry out my duty by directing you as to law and summarizing the facts. The first bit of law is to explain what I do and what you do. It is for me to state the laws, and you may take it from me as stated. The facts on the other hand are entirely for you."

Solemn, simple, astute, Hill reviewed the salient testimony. He recounted Puloka's description of the wounds. He went over everything that To'a had said and most of what Stojanovich said, as well, and the ways that factual issues bore on points of law.

"You will pay close attention to the doctors but you decide this matter. Not any doctor. The defense has suggested that the accused thought it right to kill Deborah Ann Gardner to destroy the devil. The prosecution says this is merely what the accused told Dr. Stojanovich and must be very suspect. After all, two other doctors examined the accused the night of the death and said they observed nothing peculiar about him.

"There is evidence of the accused being very worried. You will remember that Victoria Lynn Redpath gave evidence that his love life was not entirely happy, and that he was upset about a friend."

The burdens of proof were different. The Crown had to demonstrate Dennis's guilt beyond a reasonable doubt. The defense had only to show that Dennis was insane by a balance of probabilities.

The judge wrapped up in military staccato.

"If the prosecution failed to satisfy you beyond a reasonable doubt that he struck the blows, you must acquit.

"If the prosecution satisfied you beyond a reasonable doubt that he struck the blows, but you decide by a balance of the probabilities that he was insane at the time—that time that he struck the blows—that his mind was so afflicted that either he did not know what he was doing or did not know that it was wrong, you should so find and bring in a special verdict, Not Guilty Because He Was Insane at the Time When He Did the Act.

"And if you are satisfied that he struck the blows and you are not satisfied on the balance of probabilities that he was insane at the time—then the verdict must be Guilty, without any qualification."

Hill finished by reminding the jurymen of their oath before God to decide without fear or favor, and they left the courtroom at 3:14 and were out for a very long time for a Tongan jury, twenty-six minutes. Then the police led them back in a line from the side building and all the people standing out in One-Way Road under the ironwood and *'ovava* trees packed back into the courthouse, their numbers now swollen by high school students leaving classes, the new group of Peace Corps teachers, Tonga 18, and others summoned by the coconut wireless, which had informed the city a verdict was imminent.

There were never crowds in Nuku'alofa. There was a crowd in Nuku'alofa.

A person could not move forward or backward outside the courthouse doors, students were mashed up against the open windows, trying to see their former teacher's face, polite new Peace Corps trainees who were determined never to be the Ugly American were flung to the side by massive Polynesians, and the Bird Lady who had not been called as a witness and was to fly away forever in an hour also shouldered her way toward the door. Oh how she pushed to get inside, but even as she did, the crowd came pushing back at her, people spilled out the doors.

Then?

Only chaos. Raw excitement, cries, cheers, bodies stumbling upon unseen stairs. The thing was over so abruptly, something had happened, those outside had no idea, till a policeman came to the doorway at the longhouse's southwest corner and shouted out above the tumult.

"*Hao, 'oku ne hao!*"

And the words were quickly translated, by several Tongan voices, He is saved.

26

Privy Council

At five o'clock on Christmas Eve, Alice made a last round of the racks to see that all the buttons were buttoned and no coats were smushed against others and then left the Bon Marché to go to a Christmas party and as she was driving west through Lakewood heard the news on local radio. She met up with Craig at the party, and they had dinner, before going on to the Little Church on the Prairie for a midnight service, and Alice sobbed through the whole thing.

She was shocked. She had believed the case to be open and shut, Emile had told her that Dennis had done it. She had not even known that Dennis was mounting a psychiatric defense. Though Rick Nathanson had quoted the psychiatrist's testimony, his short cabled articles, which had been printed in the *Tacoma News Tribune*, had not made that strategy explicit, and Alice's friends whispered that the way the Peace Corps had allowed her to discover the news was barbarous.

By then the verdict was two days old. Word had reached Peace Corps officials in Washington late December 22. Even as Rick's report for the Associated Press had traveled thousands of miles west to Australia and boomeranged back to New York to be published on the twenty-fourth, several cables had passed between Tonga and Washington, and Russ Lynch had telephoned Sheepshead Bay to give the Privens the good news.

ALL OF US VERY HAPPY, Mary told Jack.

A few Polynesian men in the kingdom were now Deb Gardner's only remaining advocates. In a sense, Alice's daughter had become fully Tongan in death.

There had been no real debate in the jury room.

"*Puke faka tēvolo,*" said Leimoni. Sick with the devil.

"*Kovi 'atamai. 'Atamai maumau,*" said Sione. A bad mind, a broken mind.

"*Fakasesele,*" said Kepueli.

And as for Feki Moala, who did not believe any of that, who had seen the photographs of the beautiful girl and thought the defendant to be normal, just eaten by jealousy, pretending to be *'avanga*, Feki had shrewdly kept his mouth shut.

Then Hill had gaveled the trial to an end, the Bird Lady had called out Dennis's name in joy across the courthouse lawn, and he had given her a thumbs-up before the flat-bed police truck carried him out of town.

Down the King's Road. Up the hill at Hu'atolitoli. Into depression.

It came as a jolt to Dennis to be returned to the prison with the pretty name, and he did what he always did when the Tongan authorities frustrated his wishes, refused all food and water and demanded his return to town. His third hunger strike, or was it the fourth? There is a banality to repetition. Again the emaciation, again the barbed-wire ceiling to the sky, again the drinking water poured out on the apron of the tomblike ward with the draughts board incised in the concrete and toothpaste caps for players.

The Bird Lady and a whole lot of other *pālangis* left on the 4:45 flight to Fiji, but no special Privy Council session met to free Dennis. The judge, in a letter to the Prime Minister, described the verdict as Not Guilty Because He Was Insane at the Time When He Did the Act, and following the law, Judge Hill sent the case on to the King sitting with his Cabinet in Privy Council to decide "the place and mode of detention." One of the members of the Privy Council, the minister of police, had a clear understanding of what detention meant—Hu'atolitoli—and he seemed determined to get Dennis accustomed to his new home.

At first, the Americans maintained their optimism. It was rumored that the Privy Council was going to hold a special Christmas session to set Dennis free, and it certainly appeared that the royal family and the American government shared a strong interest: that Dennis be gotten off the island, quickly. So Peace Corps scheduled Dennis on one flight after another.

While in Suva, the chargé d'affaires cabled Wellington to assure the Ambassador that the United States had this one in the bag.

TONGANS FROM THE BEGINNING HAVE WISHED THAT PRIVEN-GARDNER CASE WOULD SIMPLY DISAPPEAR, THAT DENOUEMENT AND POSSIBLE PUNISHMENT COULD TAKE PLACE ELSEWHERE.

But the chargé d'affaires had spent all of eight hours in the Kingdom since Deb's death. He had not reckoned with the Tongan professionals, who were of no mind to see the case disappear.

The special session of Privy Council failed to materialize, and Judge Hill filed a report to the Stone House saying that Dennis was a "volcano" who would require "long-term, if not permanent" incarceration because of the danger that he would act in the same way again. While the Tongan director of health, Dr. Foliaki, made a report that virtually dismissed Stojanovich's conclusions: "He is fully in touch with reality—surroundings, space and time. There are no gross thought disorders, delusions or hallucinations."

As for Tevita Tupou, his report to the Stone House was more an anguished soliloquy than a legal filing.

"From the time that the murder was committed until the end of the case I found a strong Peace Corps effort, in particular by Mary George, in the defense of Priven. It appeared to me that all pity was with Priven and none was shown to the dead girl.

"The Peace Corps effort may have been made to try and save the name of the movement from the embarrassment of one of their members being convicted of murder. I find this very strange justice if this was the case, as it was another of their members who was the victim."

This belief had long been whispered among clerks, lawyers, journalists, teachers, and policemen, but no one had stated it so baldly before. Now here it was, on a paper submitted to the King, with emphasis on the directoress—

"All pity was with Priven and none was shown to the dead girl."

Mary grew more upset by the hour.

Oh how she wanted Dennis gone, wanted this horrible distraction to the Peace Corps program to end. And oh how she feared that the police might drop him off at the office. The Peace Corps did not have a brig. She did not want Dennis rolling around town.

She had a Christmas party at her house and just about hurled her guests out at ten o'clock. She began sweeping up and cleaning out ashtrays, and the volunteers felt unwelcome. Then on Sunday, the day after Christmas, she had a run-in at the Royal Chapel with Harry Bisset, the old *pālangi* who advised the police ministry, and who disliked the Peace Corps, and another confrontation with Foliaki.

"He is going to die if you keep him at Hu'atolitoli!" Mary said.

So Foliaki went out the next morning and felt Dennis's liver and spleen, and had to agree that in another day Dennis would need to be fed intravenously and intragastrically to be kept alive, and even that would be difficult. "His determination to starve himself to death is genuine although he is using this to manipulate things to his own advantage. Because his life is in our hands, I do not want to let him persist."

'Akau'ola relented. Dennis was delivered back to the jail in town, to the loving-kindnesses of the Bouchers, while the minister gathered his forces for a last stand.

Some wag at the State Department once said that a diplomat could hold his breath and stand on his head for three years in Nuku'alofa and no one in Washington would even notice. This rule now applied. The Tongans were sorting out the fate of an American government worker who had killed another American government worker, and it might be reasonable to think that the Stone House would have commanded the White House's attention for a moment or two. But Tonga was Tonga; and 7000 miles away in Washington, the case had been smothered. The Stone House would have to make do with Suva.

At the end of December, the Tongan secretary to government Dan Tufui called the chargé d'affaires and informed him of the Kingdom's baseline demands.

The Kingdom could not let Dennis go without written assurance from the U.S. State Department that he would be placed in detention. The boy's parents, too, had not come to the trial, they had not expressed remorse. They must now do as any Tongan family would and state that they were responsible for their son and would put him away whether he liked it or not.

On January 5, the Americans responded with a formal appeal to Prince Tu'ipelehake, the prime minister, to hand Dennis over. Peace Corps lawyers in Washington had drafted the letter, which had then been reviewed by a group of midlevel officials: Jack Andrews in NANEAP, the consular desk at the State Department in Washington, and the Ambassador in Wellington. And the chargé d'affaires had then affixed his flourishes and signature before sending it on to Nuku'alofa.

> Your Royal Highness:
> I understand that the Privy Council soon is to discuss disposition of the case of Mr. Dennis Priven, the American Peace Corps volunteer recently acquitted by reason of insanity of the slaying of Miss Deborah Gardner. I know that the tragedy that befell these two young people far from their homes has weighed heavily on Your Royal Highness and on ourselves.
> I have been authorized by Ambassador Selden in Wellington to convey the following information he has received from the Peace Corps for the Privy Council's consideration:
> a. In the event Mr. Priven is released to Peace Corps custody, the Peace Corps will arrange his transportation to the United States with escorts, including a medical officer."

Points b and c involved notification of airlines and limitation of Dennis's passport.

> d. In the event that Mr. Priven refuses to continue his journey at U.S. stopovers, the Peace Corps will notify local authorities. U.S. state laws would permit his apprehension by local police.

e. The Peace Corps has arranged for Mr. Priven's immediate admittance to a hospital in Washington where he will be examined by a psychiatrist.

f. All relevant jurisdictions in the United States, including Washington, New York, and Maryland, have involuntary commitment procedures. And Mr. Priven's parents would file necessary commitment papers in the event Dennis refused voluntary commitment.

g. Commitment in the United States ordinarily entails (1), confinement in a mental institution, (2), continuing psychiatric care, (3), periodic psychiatric evaluation, (4), termination of commitment upon findings that patient is no longer a threat to society or himself."

As for the Privens' promise, it arrived by aerogramme, a few short lines scratched out by hand by Sidney Priven to Prince Tu'ipelehake.

Dear Sir:

It is our hope and understanding that you are considering sending our son Dennis Priven back to the United States.

Please be assured that, in the event that Dennis does not voluntarily sign himself in to be treated, we, the family will so sign. Thank you.

Sincerely yours,

Mr. & Mrs. Sidney Priven

Peace Corps carried the aerogramme to the Stone House, and crossed its fingers as it did. No one was to tell Dennis about his father's letter. He was estranged from his parents, he wouldn't be happy to learn that they'd promised to stick him in a loony bin.

Do Not RPT Not Advise That Dennis Be Apprised This Development,

the lawyers warned Mary.

The letters angered 'Akau'ola—they were too vague. And Tufui, a courtly, sophisticated man and an experienced lawyer, also raised concerns. There were plenty of suggestions that Dennis would be committed, but no explicit assurance. The purported hospital had no name or address, and there was nothing about "detention," which Tongan law required. And Tonga would never turn Dennis over to the Peace Corps.

For once, Flanegin was thrown.

SERIOUS HITCH HAS DEVELOPED IN TURNOVER ARRANGEMENTS.

So Washington provided fresh terms, and the Stone House offices typed them up and circulated them to the Privy Council. Peace Corps was out of the picture, there was little ambiguity in the arrangements.

> Mr. Priven [is to] be handed over to Mr. Flanegin at point of departure from
> Tonga on 13 January, 1977 for repatriation to the U.S.A. for detention and
> treatment at Sibley Hospital, 5255 Loughboro Road, Washington, D.C., U.S.A.
> until Mr. Priven is declared by competent medical authority that he is no longer a
> threat to the community and himself.

It had rained off and on for the first few days of the year. Now it was the seventh and the sun made a white sheet of the wet palace verandah, and a fish fence stood out against the pale horizon like so many sharp vertical lines in india ink, and a couple of fishermen's boats made blots.

The ministers came through the French doors, seven large men in dark skirts and woven ta'ovala, and a smaller man, dry as a stick, Dr. Tapa, who had spent the week reading pulp crime books, and with a flurry from the chamberlain, an interior door opened and His Majesty, Taūfa'āhau Tupou IV, appeared, also in skirt and mat, but topped by a white military tunic. He sat at the east end of the long dark table, facing his brother, Prince Tu'ipelehake, leaner, carved by diabetes.

An old-fashioned pale green globe tilted on a high base beside His Majesty. On the walls were photographs of his ancestors, and etchings from Captain Cook's trips.

The King was 58, his ministers in their forties and fifties. They all had one foot in feudalism and the other in modernity. They had all spent time overseas, a couple were highly educated. Still, they deferred as chiefs to His Majesty, and the joke went that Privy Council had an important job, to order brooms for the palace.

"This is a very serious case—"

'Akau'ola had risen to speak, and appeared to tremble with feeling.

The police minister was unpredictable. One could never be sure which facet of his personality he would expose, the bluster, the formality, the distance, the hardness, the rage, the highflownness. Now came passion. He wanted nothing less than to overturn the verdict, and hang Dennis after all.

He spoke of Deborah Ann Gardner and her service at his alma mater. He described the nature of the attack, inside a house provided to her by the Kingdom.

Only in jail had Priven acted insane. He sang sometimes at night, and threw his own excrement. And the American psychiatrist spent more time on the witness stand, three days, than he did interviewing him.

"I will never be convinced that he was crazy. Justice means that punishment should be shared on everybody, regardless of their background. Had a Tongan committed this crime, there is no question what we would be deliberating now—only the time of his death."

All the men knew who would respond. Across from the hard womanless minister, Langi Kavaliku sat patient, smooth, warm, and wifed. He was the minister of education and works, and the most educated man in Tongan history. He had degrees from Harvard, Cambridge, and the London School of Economics, and had spent years in the United States.

"The minister of police is quite correct, this is a very serious case.

"And I tell you, we all feel responsible. I met the victim myself. She was keen, and she was dedicated, and so pretty that I wondered, What is she doing here? And then she died, and people felt they could have done more to save her. That is unrealistic, I know. Still, that is the feeling, and it was widespread. We felt that we failed this girl.

"Speaking personally, I still feel that way now. This girl was working for me. Why did I not do something to prevent this happening? A foreigner is killed, we feel sorry for her family. Her parents have never come here. So we feel even more responsibility—"

Yet Tonga had done what it could. The police had worked night and day, the Crown mounted a vigorous prosecution. And though Kavaliku agreed with the minister, that the accused was consumed by jealousy, and he could not understand how a doctor could examine someone for so short a time and render such a complex diagnosis of a condition no one else had observed, still, that was the process.

The independent process had been followed to the letter. And that process did not permit Privy Council to reverse the court's decision. Privy Council must respect the rule of law.

'Akau'ola rose again. He was a romantic who would keep on fighting though he was sure to lose. He spoke of the American obstruction, the American lies. The directoress. The *Fiji Times*. The booking of one flight after another for a savage. And all to save the name of the movement.

"How can we trust these letters—we cannot—I have seen men like this. A man like this is sure to kill again, given the opportunity. Then the finger would be pointed at Tonga."

The minister was right, Kavaliku said. Peace Corps had not behaved in a way to make itself proud. But Peace Corps was nowhere mentioned in the papers. Tonga had refused to give him to Peace Corps.

For a third time, 'Akau'ola took his stand. There was a place for Priven in Tonga. Even if Privy Council accepted the verdict, there was a building at the prison for the insane.

"Let him stay in Hu'atolitoli for a long time. On the soil where he spilled blood."

In the end, this is how Kavaliku dispatched 'Akau'ola, by invoking the central Tongan idea of *'ofa*. *'Ofa* means compassion, love, as in the translation of Peace Corps, They Work for *'Ofa*. Of course there was *'ofa* for the girl. But there should be *'ofa* for the man as well.

They had all seen American hospitals, in movies and with their own eyes, too, and American hospitals were modern facilities with the best expertise. Hu'atolitoli was in shocking condition. The minister had said that himself when he asked for more funds. The building for the insane was not the sort of place that any insane person should be held. The floor boards were loose and a person could merely lift them up and crawl out into the general yard. The Americans had promised to put him in a modern hospital, for many years, and that is where a person as disturbed and vicious as—

'Akau'ola grew silent, and ever more distant. The sea had turned to metal plate in the sun, and a small figure etched in it threw his fish net two or three times and gathered it to his waist, then wandered left toward the fine lines of the fence. At the other end of the harbor, some sort of cargo boat could be seen moving slowly, picking its way through the reef.

Somewhere far away, His Majesty spoke in a deep-voiced mumble.

"Number 4, it is resolved, once Mr. Priven is out of Tonga, he be not allowed to return to the Kingdom. . . ."

Now it was only necessary to take a vote. 'Akau'ola's face was dark, his eyes small as seeds.

All through that long Tongan spring and summer, he had contemplated his duty. The hanging tools were kept in an out building at the police compound in a box on the floor: fancy English ropes with black silk stitched around the nooses, and rusted iron manacles for the ankles and wrists. The gallows was kept at a warehouse in town. An elaborate wooden structure, it required two trucks to be carried out of town, to Mo'unga Kula.

Mo'unga Kula meant red hill. It was a hill only to a Tongan. Anywhere else this bump in the interior would be considered a mound. It rose above the red soil on the plantation land at the east end of Hu'atolitoli. Sloppy roads went back to it through the prison farm.

Ordinarily Mo'unga Kula was kept filled with sand because that was the superstition: if Mo'unga Kula was ready then someone would be sure to hang.

How often 'Akau'ola had imagined that morning . . .

The hill has been dug out. Dirt steps have been cut into the inner wall, so that after ten minutes or so, two men can go down inside, a doctor and a prison officer. A jury of doctors and ministers is gathered. They sit in chairs in the shade. The site is ringed by policemen, and outside the ring of policemen, a second and third ring of troops, the Tongan Defence Force, the Tongan army with their rusty bolted rifles. They are here to assure that no one will interfere with this solemn and sovereign task.

The sky is gray, the dreary horizon is dabbed by coconut trees. The prisoner stands on the platform. A black hood has been tugged down over his large square head.

Two officers and two prison wardens have also mounted the stairs to the little stage. 'Akau'ola is the highest-ranking. He checks the knot on the little string that ties the heavy loops of rope up in the air over the man's head. That is clever, to keep the loops of rope suspended for an instant as he falls, and prevent them from entangling him—

The minister studies the face of his watch as the seconds trickle away. He is impossible to read. None of his men is sure who is to carry out this awful duty. Each has signed a form. Each is prepared to do the job. Each has also said his prayers, that this duty might somehow pass him by.

At the stroke of 8, 'Akau'ola raises his right hand. He has handled the stiff dress, he has seen the photographs of her body. Now he steps curtly past his men to bend to the wooden shaft and throw his shoulder into it.

And afterward, a cigar and coffee.

They went around the table, the scribe's pencil scratched away.

"Baron Vaea."

"Affirmed."

"Dr. Tapa."

"Affirmed," said the doctor who had spent the week reading pulp crime books.

"Honorable Tuita."

'Akau'ola had hoped for help from Tuita, the bosslike minister of lands who disliked the Peace Corps. "Affirmed," said Tuita.

"Honorable Ve'ehala."

"Affirmed."

"Prince Tu'ipelehake."

"Affirmed."

"Doctor Kavaliku."

"Affirmed."

"Doctor Tupou."

"Affirmed."

"Honorable 'Akau'ola—"

'Akau'ola gazed off, defying His Majesty for a few last seconds. "Abstain."

27

Never-Never Land

The volunteers sometimes said that they were in never-never land. They meant that before long this island was going to vanish in a puff of smoke from a Rolls Royce airplane engine, and they would return to reality, and every untranslatable thing about the Kingdom, 'Akau'ola quoting Freud in the newspaper called the Minute of Silence, the Crown Prince with his monocle, the bicycle store that was also the German consulate, the women hammering on tapa cloth, boys firing bamboo cannons with kerosene, men with machetes sitting atop a whale in the harbor, the transvestites having a beauty contest, the *fokisis* on the handlebars, everything would slip away beneath the clouds.

The volunteers had some idea of what would happen when they got back. "America doesn't exist when you're here. Tonga doesn't exist when you're back," older volunteers had told them. In no time they would pick up their lives. Their friends would lose interest in their stories, and the Kingdom would be as foreign as a hot shower was now. Then one day they'd open a drawer to discover something they had brought home, a tortoise shell bracelet, or a folded piece of tapa, and they would wonder, Did that really happen? What did it mean?

On Thursday, January 13, 1977, at four o'clock, a police guard brought Dennis to Fua'amotu airport and after instructing him that he was not to return to the Kingdom—on pain of what?—turned him over to a silver-haired American in a blue blazer.

The chargé d'affaires had come in that morning for another lightning visit, and found it necessary to make some adjustments to the receipt the prime minister's office presented, reflecting last-minute concerns from the consular desk in Washington.

DEPARTMENT POLICY PROHIBITS RPT PROHIBITS DIPLOMATIC AND CONSULAR PERSONNEL FROM TAKING PHYSICAL CUSTODY OF AMCITS IN SITUATIONS AS IN THIS CASE. OFFICERS MAY HOWEVER BE PRESENT TO FACILITATE THE RELEASE OR REPATRIATION OF AN INDIVIDUAL AS LONG AS IT CLEAR THAT SAID OFFICER IS NOT TAKING PHYSICAL CUSTODY OR LEGAL RESPONSIBILITY.

So Robert L. Flanegin scratched out the word "custody" on the receipt, and signed it, then brought Dennis on the plane. As the chargé had stepped on to the plane with Deb's body, after his last visit three months before.

As this point one of the necessary fictions of the arrangements dissolved: Dennis wasn't being handed over to the Peace Corps. Paul and Laura Boucher were on one side of him, because they were his friends, yes, but because Paul was thought to be big enough to restrain Dennis, and next to them was the new Peace Corps doctor, Clark Richardson, who'd worked for years on tropical medicine in Vietnam, and down the aisle, incognito, Qantas having insisted on it, an airline security man who looked back at Dr. Richardson every fifteen minutes and got a little wink.

The chargé peeled off in Fiji. Then there were four more flights, the last of them a TWA red-eye to Washington. They sat in VIP lounges between flights and Dr. Richardson made a point of staying awake the whole way, but the Bouchers went in and out and Dennis was zonked most of the time.

"So you don't get flight-sick, take a couple of these—Dramamine."

The doctor proffered a palm full of an antipsychotic that Dennis ate all the way to Washington, this being standard procedure for a Peace Corps psych-evac. "Disturbed psychotics" were to be accompanied and medicated, with a straitjacket on hand, and delivered as quickly as possible to Sibley Memorial Hospital.

They got to Washington pushing ten in the morning. The Bouchers went to the hotel, and Dennis went on with the chief Peace Corps medical officer to Sibley, where he refused to go in.

This set off a panic. Dr. Rosa called the counsel's office. Barkow told him to bring Dennis to headquarters, then Barkow called the Bouchers at the hotel and asked them to please come in, too.

Killer and doctor stepped out of a car at headquarters amid the banging of hammers. The city was making ready for the inauguration of Jimmy Carter six days hence. Stands were being erected, bunting tacked.

Dennis demanded an airplane ticket back to New York immediately.

"I explained there would be some delay in getting the ticket and we discussed the situation for about an hour," Barkow wrote, in another of the endless memoranda the counsel's office created in the case, as if this was all by the book. "He stated that he did not wish to be placed in a locked ward because of his recent incarceration, he needed some freedom. . . . He would give some thought to returning to the hospital at a later time."

It was a tense situation. Dennis got bullish and insistent, his friends had to keep him calm, and Paul Magid called the Washington city police to say they had a killer on the premises, what was to be done.

Nothing, said the D.C. police. "Unless Dennis committed an overt act which demonstrated he was an immediate danger to himself or to society, the police could not take him into custody," Magid wrote, for the files. "The fact that Dennis had been acquitted of murder by reason of insanity in another jurisdiction did not provide a sufficient basis upon which to make this finding."

Peace Corps and the State Department had promised the Tongans that in the event that Dennis walked away, the American government would notify local authorities, and "U.S. state laws would permit his apprehension by local police."

That was a misrepresentation. No one on the Tongan Privy Council was a lawyer (though His Majesty had an Australian law degree, he had never practiced), no one on Privy Council understood what any lawyer could have told them: the American letters had no legal power. Without a treaty between the two countries to govern the exchange of prisoners, Tonga could do nothing to enforce the American promises. The State Department had actually given thought to establishing such a treaty to deal with the case. But the thought had been based on the prospect of a conviction, if Dennis were sentenced to hang or rot in prison.

Establishing such a treaty "will be based on need," a State Department officer said, and the South Pacific "would have low priority."

Well, of course it had low priority. No one in the United States knew what was going on. This was the essential difference between the two countries' responses. Every conceivable official in a microstate had been engaged by the matter. Meantime in the United States, a few middle managers were the only people aware of the case. The letters they'd handed over were mere assurances from one government to another, to which neither the Gardner family nor Dennis Priven was a party.

Privy Council had not been let in on the joke. The ministers had understood the letters to be legal documents. Kavaliku had pointed out that three different commitment laws applied, in three different states, and Dennis would be held till he was no longer a danger.

"I believe that under these circumstances the insane are held for ten years or more, sometimes all their lives," Kavaliku said.

Other promises in the letters were misrepresentations. Dennis's parents could not compel his hospitalization; Dennis was a grown man. Sibley Hospital was not a place in which an insane person could be incarcerated against his will till he was no longer a danger. Peace Corps said Sibley because Sibley was the hospital to which it sent all its psych-evacs. But Sibley was a private hospital. To get into the psych ward, a person had to check himself in.

Even the chargé d'affaires had been aware of that. As he told the consular desk at State,

PEACE CORPS HAS TENTATIVE PLANS ENTER PRIVEN IN WASHINGTON HOSPITAL, ASSUMING HIS APPROVAL.

"Assuming his approval" is not what the chargé had written to the Tongans.

By the time Dennis was not granting his approval in Washington, though, Flanegin was back at his desk, half a world away, announcing the completion of his mission to the Ambassador with restrained grandeur:

"Three months from the day we learned of the arrest of PCV Dennis Priven for the slaying of PCV Deborah Gardner in Tonga, Priven was released from jail by order of the Tongan Privy Council for repatriation and treatment in Washington. The case was protracted and difficult but I believe the outcome was the best possible for all concerned."

He had spent all of 16 hours in the Kingdom on the case, and the Ambassador, who'd spent not an hour, clapped him on the back.

BELIEVE YOU HAVE DONE A VERY GOOD JOB IN THIS COMPLEX CASE.

The Peace Corps had one small power to hold over Dennis. He was worried about his future.

Sitting in a VIP lounge in Fiji, Dennis had mused in his detached way about whether this whole thing was going to have an impact on his career, and the line had gone rippling back through Peace Corps in the form of gallows humor. "What do you say to a murderer who asks, 'Is this going to look good on my résumé?'" And really that was the only leverage that Peace Corps had left.

At three in the afternoon that Friday, the associate counsel met with the NANEAP director and upon consulting the Peace Corps director, they came up with a plan. Peace Corps would hand Dennis two letters that afternoon. One would authorize his travel to New York. The other would be a medical termination of service. Medical termination would be a blot on Dennis's record. The plan seemingly worked. Dennis didn't want a medical termination, and he agreed to stay on in Washington that night at the National Hotel on I Street.

The doctor was to monitor him. Dr. Richardson was in the next room with a key to Dennis's room, and his ear to the wall.

Monitor Dennis, make sure he doesn't run away.

The Tongans had always felt that the CIA was in on the operation, but even the CIA had never pulled off a caper like this. Richardson soon fell asleep. He could be excused, he'd been up for two and a half days. Then when he got up, it was a sunny Saturday morning, and panicking, he ran into the next room and found the prisoner gone. He'd failed his one responsibility, and flew downstairs to find Dennis eating breakfast with his mother in the little restaurant.

Again things were tense. Dennis didn't get on with his parents, but he was determined to go back to New York. Several impromptu meetings were held. Then Dennis's parents headed home, and Dennis agreed to go into Sibley to be assessed.

In the months to come, back in Tonga, Paul Boucher would go silent or angry when volunteers and students asked him about the case. He'd shrug off the question, or say, "It doesn't involve you, it's not any of your business." But on that Saturday morning in Washington, it would seem that it was Paul and Laura who served the public interest in the case, and compelled their friend to check into the hospital.

From the moment that Paul had set off on his bicycle with Dennis to go to the police station on the night of the murder, the case had drawn him and his wife into a highly unusual situation, with all sorts of ramifications, geographical, psychological, legal, and spiritual. The Bouchers had done all they could for Dennis. They had visited him and fed him, and Laura had fought for him.

They had been Dennis's American hosts outside of American territory. Notwithstanding the old Peace Corps principle that a volunteer was to be judged in local courts by local judges, the Bouchers and Dennis's other close friends believed that Dennis was not well served by Tongan justice; he needed help. "My sympathy lies with Dennis who obviously needs psychological help and this he will not get here," Gay Roberts wrote. As the Bird Lady sat listening at the edge of the circle in the jail, Dennis had told the Pauls something of his actions on the night of October 14. Then later on, during a discussion among the friends at the Bouchers' house, she had learned more about Dennis's feelings: that Dennis felt that Deb deserved to be killed because she had been intimate with men and then thrown them away.

The friends had kept their mouths shut as Crown investigators swirled around them trying to make a case. The friends had lived in a climate of fear and secrecy. None of them had gone forward. But none of them had been directly questioned either.

Except Paul Boucher.

Paul had met with Clive Edwards and Kosta Stojanovich extensively before the trial, then Paul had worn a suit to testify and under defense questioning, he had laid the grounds for the insanity defense. He was a close friend to Dennis. He'd visited Dennis's house once or twice a week, Dennis had visited his four or five times a week, and had dinner over there once or twice a week.

Paul described dramatic changes in Dennis's behavior, beginning about the time that Dennis painted a queer face on his door.

"In about August he began to change in many ways. He became more and more depressed and very frustrated. Various things were frustrating him. He hoped to stay in Tonga for another year. I think he was turned down. He felt that no one was paying attention to his request. He became very bitter. He was not acting normally. Small things upset him excessively. He seemed to magnify things."

Judge Hill scratched long lines alongside Paul's testimony, and wrote, "Defence Case."

And then Tevita Tupou got up and specifically asked Paul about the relationship between Dennis and Deb.

"Did he talk about Deborah?"

"He had mentioned her name. He spoke little about her."

"Was he frustrated with her?"

"I can't remember him saying he was frustrated with her."

"Did they have a relationship?"

"I'm not aware he had much of a relationship with her, I knew they were friends."

These statements are hard to believe. Paul Boucher was a highly discreet person, with a legalistic bent—and he could justify his answers because he felt that the prosecution was asking him what Dennis had said about Deb before the murder. Before the murder, Paul and Dennis had never talked much about girls. All the same, many people knew that Dennis was hung up on Deb. Emile Hons knew it, Paul Zenker knew it, Vickie Redpath knew it, and so did the Bird Lady. Peace

Corps lawyers back in Washington knew it. "Debbie's unresponsiveness was a great source of frustration to him," said a lawyer's memo four days after the murder. Frank Bevacqua had known it since the failed dinner party of the previous July—a dinner party that Paul and Laura had helped Dennis put together.

The Crown wanted that kind of information, and it tried to get it from the man Dennis considered his closest friend. And Paul said that Dennis didn't have strong feelings about Deb.

Now in Washington, it was Paul and Laura who urged Dennis to do what the close friends had felt he should do back in Tonga: get help. If anyone could make Dennis go into the hospital, it was the Bouchers. Later, when they got back to Tonga, the Bouchers told Paul Zenker about the scene at headquarters, and he recorded it in his journal: "Paul & Laura said that Dennis was not going to go into hospital in DC despite their pleadings, so they left him. Their take: the impact of them leaving, Laura crying, and Dennis' parents coming in led him to go in. Dennis' mom's first words to him were, 'Why did you do it Dennis?'"

So Dennis went into Sibley to be assessed, and got a plastic hospital bracelet, and came and went.

Big, charismatic, kindly, and lordly, with thick glasses on his big nose and a soothing voice, Zigmond Lebensohn, 66, was chief psychiatrist at Sibley and a grand presence in psychiatry. He had worked at the National Hospital in London, he'd advised Ann Landers. A lot bigger than the first psych, but not as quiet or burning.

Lebensohn spent hours talking to the patient, and the case made perfect sense. Dennis Irving Priven was a shy and withdrawn man with very little sexual experience, the girl was the kind of girl he had only been able to dream about back in the States, beautiful and kind and attentive, and here she was, in his life on a little island.

She had led him along then slammed the door, she'd gone out with several men but not with Dennis. It was a kind of taunt and it set Dennis off.

"For this kind of guy, that triggered everything. Everything went kaflooey."

Had Lebensohn judged Dennis to be dangerous to himself or others, he could have phoned the police and had them arrest Dennis as he walked out the front door and take him on to the government hospital, St. Elizabeth's. But the psychiatrist determined that there was no need to do so. Yes at times Dennis lost his temper and grew belligerent, but he was after all in a frustrating situation. Dennis didn't strike Lebensohn as psychotic or even highly disturbed.

The boy was essentially tractable, he'd suffered a bad break, maybe a psychotic break, but he could recover. He wasn't a paranoid schizophrenic, he didn't have megalomaniacal delusions. "He said, 'I am her Jesus Christ,'" Dr. Stojanovich said. But Dr. Lebensohn heard no such talk. Nothing about the devil, or the savior either.

And the fog?

"He had a keen memory. He remembered the experience very well, he didn't want to talk about it."

If Lebensohn had been testifying in Tonga, Dennis would not likely have met the requirements of the insanity law: he didn't have a mental disease that kept him from knowing what he was doing or knowing that it was wrong. As it was, Lebensohn gave Dennis a prescription for a sedative and the name of a psychiatrist in New York who it was important for him to see, and Dennis said he would see him.

Peace Corps staff were afraid of Dennis, but Lebensohn calmed them down. They had nothing to fear. Dennis had a "situational psychosis," his violent reaction had been very specific to a situation. They weren't a woman leading Dennis on, they shouldn't worry—and one of the lawyers organized a dinner at his house so that they could keep an eye on Dennis.

Now and then Dennis showed up at headquarters to pick up his wage at the cashier. He was still on the books as a volunteer in good standing, he'd outlasted just about his whole group. He rode the crowded elevator down to the street, pale and quiet as could be.

Brad Dude, an associate country director in Samoa, saw Dennis on the cashier's line and ran over to NANEAP. "What the heck is going on? I just saw the Tonga killer upstairs."

"There's nothing we can do to hold him," said someone on Jack Andrews's staff. "Don't advertise it, OK?"

When Dottie Rayburn, the country desk officer, had to see Dennis, she asked the chief budget officer to keep an eye on her desk. Dennis sat there fulminating about Mary George, what a bad country director she was. Till finally Dottie pointed out, "You'd be rotting on the prison farm if not for her."

Someone had to tell Balzano. A couple of lawyers went to the Action director's door.

"That volunteer who killed the other volunteer in Tonga—" Russ Lynch said. "Well, it looks like he's going to have to go free. He's killed somebody, but he didn't commit a crime here. He's not guilty of breaking any law here."

Mike Balzano flew off the handle.

"What! What the hell did you guys do? You mean we can't get him for anything? What about destruction of government property? She worked for us too."

The lawyers didn't think much of that angle, though Balzano's counselor, Harry Hogan, said that when he'd been attorney general in Oregon, a guy had beheaded a woman with an axe and Hogan had seen to it that the man would never get out of the insane asylum, and if he did, he could be tried for murder.

Well, that was so much useless anecdote. Balzano got a sick feeling, and saw a sick feeling in Russ Lynch's eyes, too.

"His own uncle says the family's afraid of him," Lynch said. "He's a sick boy and he shouldn't be going free, he should be staying in Tonga."

"Russ, we've just let a loaded rifle walk into the goddamned city."

Not that he did much about it. Balzano was leaving office in two or three days, making way for a progressive Democrat. And though he was bureaucratically responsible for Peace Corps, he had always and accurately understood himself to be an outsider to the program.

The one thing Balzano did was write a letter to his successor, Sam Brown, saying volunteers' lives were at risk and he wouldn't learn about it for months after the fact. The people who worked for Peace Corps were good people, morally and spiritually motivated, but "one of the consequences of this sense of spiritualism is the elevation of the priority 'Peace Corps must survive' above all other considerations."

This meant that Peace Corps would always put a higher priority on the life of the program in a country than it did on the health, safety, and welfare of volunteers. And the State Department would play its part, applying pressure to maintain an American presence.

"In this letter I have purposely avoided specifics. However I have a moral obligation to the present and future volunteers to convey to you what are clearly signs of danger. God bless you and everyone who serves under you."

Dennis was leaving Washington along with just about everybody else in the Ford administration. Across Lafayette Park, reporters swarmed the last White House press conference to say farewell to the benign president who had pulled the country out of the Watergate ditch.

"Last night after the President and Mrs. Ford returned from dinner, there was a surprise party by members of the president's staff, the Cabinet and friends of the Fords on the State floor of the White House—"

"How many?"

"I don't know, there were probably two hundred people there."

"It was a surprise for the President?"

"No, it was a surprise for Mrs. Ford."

"Why?"

"She didn't know about it ahead of time—"

Laughter.

"Why did they want to surprise her?"

"The President wanted to surprise his wife."

"Ron, was anything said by anyone? Did anyone make a little speech?"

"No it was just a pure—"

"How long did it last?"

"I don't know what time the last dancer left."

The only reporters who'd followed Deborah Gardner's story were upstairs in Tūngi Arcade in Nuku'alofa, working for the government. The *Chronicle* had published the last news in Tonga about Dennis Priven on Thursday, January 13, 1977, the same day that Dennis was released to be hospitalized in the United States. Paua had telexed the item to Sydney, and the Associated Press had put it out in the United States, also to appear January 13. That was the end of it. Dennis's and Deb's names would not appear in print for another quarter century.

Back in never-never land, Mary called the palace to seek an audience with the King.

The audience was granted, and so it was the dance of the orange soda, waiting for the King to lift his glass to his lips and signal that the meeting was over, and meantime Mary thanked His Majesty for the just, humane, and expeditious way that he had dealt with the tragedy.

"I hope you still see value in the Peace Corps," Mary said.

The King said he still wanted the Peace Corps around, but henceforth he would appreciate it if Peace Corps did psychiatric screening of all volunteers who were chosen to come to the Kingdom.

Mary said she could understand his concern but hundreds of volunteers had served in Tonga without incident. "And Dennis Priven had served as a successful volunteer for two years and was just about to come home prior to the tragedy."

After that Mary sent off a memo to Jack Andrews, urging Peace Corps to go along with the King's request. The program had been saved, things were generally upbeat, but the case had had a giant impact, and there was still "some negativism and hostility evident" in certain parts of the community.

"I shall convey to you, personally, my heartfelt gratitude for your support when I have the pleasure of seeing you again," Mary signed off.

"Best love to all of you in NANEAP."

Jack sent Mary's memo on to Dellenback on the twelfth floor. The nonpartisan former congressman had managed to hang on into the Carter administration, and Dellenback wrote to His Majesty to assure him that everybody who applied to be a volunteer was screened to the greatest extent possible and that "Any applicant with discernible psychiatric problems is disqualified from Peace Corps service."

That was another misrepresentation. Peace Corps had dropped psychiatric screening. Dr. Lebensohn was fighting to have it reinstated.

As for Dennis, Dellenback said that he was one of those volunteers who had a "completely normal" history till the "cultural change" of service had precipitated mental illness.

"You will be interested to know that Peace Corps Volunteer Dennis Priven was returned to the United States without incident. He entered Sibley Hospital for psychiatric care and he has now moved to New York, where he resides, for further treatment. We understand he is making good progress."

Not that anyone was keeping tabs. The Tongans had been promised that if involuntary commitment procedures didn't work in Washington, D.C., the American government could compel Dennis's hospitalization in two other jurisdictions, Maryland or New York. But Peace Corps didn't push it, Peace Corps breathed a sigh of relief. Dennis was now a private citizen. He'd left Washington some days after he arrived (three days or twelve, by varying report), and Peace Corps paid him through the end of January. After threatening to give him a medical termination, it gave him a clean record in his public file: Dennis Priven was "COS". Completion of Service. "Returned Peace Corps Volunteer. Clear. No Qualifier," it added. So nothing happened.

Dennis wasn't done. Soon after he got back to New York, he applied for a new passport. The Peace Corps had lifted Dennis's special no-fee passport when he came home, now the State Department called the Peace Corps and said, Is there anything to prevent him getting a passport, any court order? And Peace Corps said, No, there was no court order. So four months after Deb Gardner had been stabbed twenty-two times, a new passport was issued. And once again, the general counsel's office documented it, in a ho-hum memorandum.

Peace Corps had tied a ribbon on the case.

"It appears to me that Peace Corps has finally finished with the affair of Priven," Barkow wrote to Dr. Stojanovich in Hawaii, catching the doctor up, while Dellenback used the same word in a letter to his former congressional colleague, the Ambassador in Wellington. "You and your people really have been most supportive in a series of matters, Armistead, including this Tonga affair."

Tongans are so utterly provincial that they are not at all provincial. They know better than citizens of a superpower where they stand in the world's estimation:

they are a dot on the globe. And though they struggle for scraps of recognition and resent being overlooked and maligned, still, the Tongans know their place. We are nowhere. *Fo'i 'one'one*, they call their islands—single grains of sand.

The King did not let this Tonga affair pass without his own sly signal to the Americans. He sent a big box to Wellington with a stiff card: "From King Tāufaʻāhau Tupou IV and Queen Mataʻaho for your Excellency and Mary Jane Selden." Inside was a carton of green bananas and two loose watermelons. Exactly what the Americans would expect from never-never land.

28

The Earthquake

The story of what had happened in Washington soon got back to the Kingdom.

Headquarters had advised those who knew about the matter not to talk about it. The Peace Corps's desire to keep the thing quiet was now wedded to a firm legal principle: Dennis is a private citizen, we cannot talk about his medical treatment. He was like any other psych-evac, and Peace Corps didn't talk about psych-evacs.

But the story from Washington was simply too shocking for people to keep to themselves. Doc Richardson was discreet but he was flabbergasted and he blurted two words to Mike Basile, "He walked." Mike was also flabbergasted, and soon the news reached Christine Oldroyd, who was appalled but not surprised, and Paul Zenker, who was stunned. Frank heard it and couldn't believe it, and he told Rick Nathanson, who by then was deep into a new story for the best little newspaper in the South Pacific and dismissed it as so outrageous and inconceivable that it simply couldn't be true. Well, he was 23. Rob Beaver, Deb's former department head at Tonga High, heard it and said it was an injustice and if Deb had been a Kiwi he would get the news into the Auckland papers. But what did he know about American papers.

Judge Hill heard it and got upset with himself. The judge was now well on the way to "going troppo," sometimes his blue eyes seemed to swim in gin. "I should have seen him flogged before he left," he said.

It was only inevitable that the news came to the man who'd cared most about the case.

'Akau'ola was enraged. He called up the chargé d'affaires in Suva. He'd heard that Priven had been released from the mental institution in the United States, was this true? Flanegin said he didn't know but he would look into it, then he cabled Peace Corps/Washington, and Peace Corps cabled him back a week or two later.

"As Peace Corps has informed the King in a letter, Mr. Priven entered Sibley Hospital for psychiatric care upon his return to the United States. He later moved to New York, where he resides, and we understand that he is making good progress."

It was the habit of the chargé d'affaires to keep his boss the Ambassador informed of doings in the South Pacific with weekly reports of a diverting nature. A shark attack, Marlon Brando's latest visit, the King of Tonga's poorly planned trip to Guam—the chargé passed it all along to Wellington. But no cable or tidbit about 'Akau'ola's angry phone call or Dennis's release went out to Wellington now. The case was done with.

The Ambassador had congratulated the chargé, the chargé had congratulated himself back. He passed the news along to 'Akau'ola and closed the book on the case.

'Akau'ola burned. He felt bitter, and vindicated. The Tongans had wrestled with wordless Dennis in the dark for three months. They knew his holds, they'd warned the Americans. But the Americans had trampled on Tongan sovereignty, in league with that brutal killer, and now he was manipulating the American system.

It was 'Akau'ola's theory that Dennis had studied the justice system like a gameboard, and figured out how to crack it. What a mind that Dennis had! Dennis had coldly schemed his way out of the situation, and used others to do so. Plainly different from the he-just-went-kaflooey theory endorsed by Dr. Lebensohn.

In this belief, 'Akau'ola had an unlikely comrade: Clive Edwards. Dennis's lawyer had gotten a shock in the last few moments of the trial. Not from the verdict itself, from his client. Dennis had turned to him with a broad smile and said, "Thank you very much," and offered Clive his hand. Till that moment, Clive had believed that his client was crazy. He had argued the case fervently in that sincere belief. Now who was this? The man appeared to him to be rational and completely aware of everything that was going on, and Clive could not get the man's look out of his mind. He wondered if Dennis had not manipulated him to believe something that was not true so that he would be more effective as an advocate.

'Akau'ola never did go public with his feelings of betrayal. He had too much pride in Tonga to go public. Though occasionally he uttered angry words about Peace Corps in Parliament. And if he trusted someone, 'Akau'ola would speak about the case to him or her. For a time there was a former Peace Corps volunteer at Hu'atolitoli on marijuana charges, an Iowa farm boy who had accepted his punishment, and 'Akau'ola decanted his bitterness to Merle as they walked around the prison farm. But generally he stored the humiliation in his own large chest.

For a while, Mary kept working, but it was like the cartoon character who runs off the cliff and keeps pinwheeling his legs, any minute she was going to drop.

Peace Corps/Tonga had been saved, and now Mary was busy busy busy. So busy that as she pressed the staff to finish up the annual country plan so that NANEAP would learn about all the wonderful new projects for Tonga, she instructed the volunteers not to come by her office unnecessarily for three months, from January to April.

"We would appreciate your cooperation in keeping your personal requests down to a minimum. Therefore I will be available to sign papers or discuss things with you only between the hours of 5–5:30, Monday–Friday, but no other time."

A half-hour a day for volunteers—that was no way to keep them on your side.

The horrible distractions of 1976 were over. Now everyone was supposed to put the murder out of mind. But the knowledge of it crept down through Peace Corps/Tonga like rot, and though Tevita's last words to Privy Council were never published, a curse does not need to be published, it need only be whispered and believed, and this one was believed.

It appeared to me that all pity was with Priven and none was shown to the dead girl.

One day Ralph "Lolo" Masi, a volunteer who had helped clean out Deb's *fale*, went by Mary's house to interview her for the volunteer newsletter, and when Mary said something about the murder, Lolo wrote it down but Mary said she didn't want him to write it down and grabbed for the page in his lap. Then it was a tug of war. Mary was holding the top half, Lolo the bottom, till the page tore, and

Lolo ran out of Mary's house with his half and Mary yelling after him, and Lolo afraid he was going to be thrown out of Peace Corps.

The staff grew restless, too. The Tongan staff said that they were overworked on the country plan, and Mary gave Mike a negative review, and Dr. Richardson wrote to Jack Andrews in outrage. "We are not saying that Mary is altogether a bad girl. We are saying that her administrative, diplomatic and leadership skills are not adequate for Tonga."

The volunteers heard that Mary wanted to ship Dr. Richardson out, and they mutinied. One wrote to NANEAP to say that the volunteers had experienced "frequent intimidation" from Mary, resulting in fear, distrust, and frustration, and they would strike if Mary was not removed.

They made a long list of criticisms. She couldn't communicate. She was too flamboyant in her dress. She gave out religious tracts. She described a vision she had when she woke up and Jesus was standing at the footboard.

Jesus came to visit Mary in Tonga.

Some of the volunteers' complaints were petty, even mean. Mary couldn't relate to volunteers, she wore long white gloves at official functions, pulled a drawer out in her kitchen and yelled when she saw a cockroach and didn't think what volunteers had to deal with, rats chewing the calluses on the soles of their feet at night. Makeup. Raid. Gushy. Withering. That out-of-my-way handshake.

And under it all, Debbie Gardner Debbie Gardner Debbie Gardner.

Of course, Washington had heard no end of muttering about Mary, even Priven had complained that she was incompetent! Still, a volunteer strike was a problem no one could finesse, and Jack Andrews sent out a deputy, Dick Hailer, to look into the matter.

Hailer met a bunch of volunteers in the upstairs of Viliami Kapukapu's house, the same room with the cushions on the floor where Deb had been called a bike.

A volunteer stood to say that volunteers didn't trust the country director. "We feel threatened by her."

"I'm glad you gave me your opinion," Hailer said. "I do not want to discuss that."

"Well, then, why did you come to Tonga?"

"The staff has been told that if they can't get it together, something will change."

"Is removing a staff member a possibility?"

"Yes, it is a possibility."

"We heard that Mike Basile got a bad evaluation. If Mike is removed it would be a real crime."

Others cried out, "Yes," and *"Poupou!"* Tongan for right-on.

When Hailer got back to Washington, he told Jack that Mary was unsuited to the job and she'd done Mike Basile an injustice in the negative review. But a month passed, and Mike wrote to Hailer to say the program was at risk. "What price do you want to pay to see her survive here? Mary is intelligent, witty, charming and gracious, in the short term. But she kills people with kindness. Or with anger, reproof, humiliation, scolding."

Nothing happened. Jack was corporate, Jack believed in chain of command, Jack had told the Peace Corps director that Mary was doing superbly.

So another month passed, and NANEAP announced its annual conference in the Philippines. All the country directors would be there, Jack would be there too. He'd demonstrated catlike political survival skills. Having landed his job in a Republican administration, he'd stayed on into a Democratic one, perhaps in part because his wife had worked for Jimmy Carter's campaign. He'd even outlasted Dellenback. Carter had named a new Peace Corps director, Carolyn Payton, and she was coming out to the Philippines too.

Mike wrote Jack a letter with "Eyes Only" at the top. A person maybe writes a letter to the boss like this once in his life, most of us never do.

"I believe the situation in Tonga to have been serious for many months. I have attempted to pass this information back to you since last December through channels. You have had at least that time, and really more, to have given serious consideration to the behavior and competence of Mary George.

"You have decided to continue her presence in Tonga as country director. That decision is a mistake. I believe this mistake is serious enough to have decided to take the following action."

First, Mike would go over Jack's head inside Action to describe Mary's unsuitability, then if Mary wasn't gone by October 14, he would go to Congress.

"If she is still not removed, I will go public. I owe it to this staff, to the volunteers, to this program and to the Government of Tonga to provide this information to those who will act decisively."

The letter didn't say anything about Deb, Deb was just the elephant in the room. Mike's deadline, October 14, was the one-year anniversary of her death.

A new member of the staff, an Episcopal minister, flew out from Tonga with the letter. It was a fabulous retreat. Los Banos is set in the forest an hour south of Manila. All the country directors and staff were in little bungalows, and the refrigerators were stocked with San Miguel beer.

The Rev. Bowen gave the letter to Jack, and Jack read it and banged the desk. "So. Basile is throwing down the gauntlet."

Jack took Mary aside in the lunchroom, and the two sat hunched and serious. Another country director came over to sit down, Jack gave her a tight smile.

"Can you please excuse us, Nancy, we have business to discuss."

Then Mary resigned.

She didn't go easily. Did she do anything with ease? No, Mary was too troubled a person, too high-flying and tremulous to do things easily. First she got sick on the plane back to Tonga. Then when she did leave Nuku'alofa, a month or so later, she didn't take much stuff and told the Tongans she was on medical leave. "I'll be back."

Mary had once told a staff member that if she were forced to resign she'd fight it through the federal courts. Wellington would have to do for now. She went to the Ambassador. The Ambassador was caught up in the protests over a visiting American nuclear submarine. "My what a silly fuss that was over the sub!" Mary said to the Ambassador, and meanwhile all her friends started writing him to say why Mary should hold the Tonga job.

Dan Tufui, the secretary to government, wrote to say that Mary was first-rate. "Physical beauty is a hollow asset unless—as in Mary's case, it is a reflection of that innate kindness and concern for others which is true beauty of character," wrote Mrs. Farquhar Matheson, the writer who had attended Mary during the trial, and Mrs. Farquhar Matheson also said that Mary had been undermined by incompetent troublemakers, a point echoed by a couple of volunteer supporters of Mary who wrote to attack Mike and Dr. Richardson.

Vickie Redpath was back in the Blue Mountains. She sent Mary a letter urging her to reconsider. "I know how difficult it is to effect change in Tonga and what a struggle it can be trying to surmount difficulties while striving to keep your faculties intact. . . . Do take good care of your dear self. God bless."

Mary sent Vickie's letter to the Ambassador with a little note suggesting that she now had Deb Gardner's endorsement.

"This letter just came from a volunteer who was here during the murder and trial. She was the best friend of the murdered girl."

The Ambassador sent the letters to Washington, but Peace Corps had moved on. It had named a new country director for Tonga, a roughrider anthropologist from New Mexico called McCrossen. He moved into the director's house with his wife, a former volunteer, and their children.

Still, Mary wasn't done.

She came back to Tonga a few months later, in 1978, at the same time as Jack was visiting the Kingdom. They both stayed at the International Dateline, and dined together, and traveled on to Suva together, to a gathering of Peace Corps directors. Jack even asked the new director, McCrossen, to make dinner reservations for him and Mary.

Well, that is insane, McCrossen thought, still he complied, and then lost it two days later when he found Jack and Mary eating breakfast together with a couple of other Peace Corps officials, and Jack offered McCrossen what McCrossen considered a white lie about how he and Mary had come to be at the same table that morning.

The small white lie "ignited a small white fire within my soul," McCrossen said. He blew up at Jack, told him he was having enough trouble convincing people in Tonga he was the country director without Mary coming back and Jack lavishing attention on her. Jack didn't take kindly to insubordination even when it came out of a person's soul. He said Mary was a personal friend. The attention he paid was his personal business and he felt no need to apologize to anyone.

But it didn't stay his personal business. The top aide to Carolyn Payton, the new Peace Corps director, soon received two letters saying that Jack and Mary had an inappropriate relationship, particularly inasmuch as her replacement was struggling to establish himself.

"Jack is a highly capable and competent person, an honest and sincere individual," wrote McCrossen. "I also think he is a human being and as subject to manipulation by unscrupulous people as any of us."

Dr. Richardson was blunter. "All last year we were wondering who was protecting Mary? Jack probably genuinely felt he could share the companionship of Mary without hurting anyone. But there is no private life in Tonga, people invariably find out what is going on.

"Mary can go wherever she wants, even back to Tonga. We cannot touch her. Jack can. Jack can carry on his private life—outside of Tonga."

Jack denied the rumors. Mary required close attention, she was a touchy-feely person, it was preposterous to suggest they were anything more than friends. But by then Jack was ready to go. He didn't care for Carolyn Payton, their styles clashed, and a month or two after the letters landed in Washington, Jack resigned.

Mary went over the horizon. Her next port-of-call was England. She met a well-off man there, and married him. It was a very religious wedding, people were standing up, crying Praise God. But the marriage lasted about ten seconds, before Mary moved on like the wind.

In the time between when Dennis left and Mary left, one other thing happened in the Kingdom of Tonga—the earthquake.

A little after 1 A.M. one night in June 1977, the biggest earthquake in recorded history hit the Kingdom, 7.2 on the Richter scale. Refrigerators walked across floors and spilled their contents. Old coral Peace Corps toilets split in half. Beds danced across bedrooms with people still on them, eyes stapled open. A mother rushed out of her house, forgetting her children, she was so scared, and coconut trees whipped back and forth catapulting coconuts through *faletonga*s.

It went on for a few staggering minutes, Francis G. Lundy thought it was never going to end. He ran out of his house to see electric wires flashing and snapping. Another Tonga 16 saw bright domes of static like haystacks on the ground. Rob Beaver brought his family out to a table in the yard and said, Each of you hold a leg when the tsunami comes, we might float away.

There was no tsunami. After the earthquake came a chatter, a clatter, a crackling. Like gunfire, mounting in all directions.

But it wasn't guns, it was people banging pots and pans, hitting sticks against metal roofs and tree trunks. That was custom. In Polynesian mythology, Maui was the god who lay under the earth, supporting it. When he turned to try and find a better position, he made an earthquake, and the people rattled their roofs to tell Maui to lie still. A3Z didn't have to put out a call to remind Tongans to do that, it was ancient and earnest.

The sun came up, and the disaster was fully revealed.

Many of the places where Deb Gardner's Tongan life and death had unfolded were damaged. The Peace Corps office had cracks in its concrete stilts that would take six months to fix. The foundation of the Stone House, where Prince Tu'ipelehake and Dan Tufui had given comfort to the American chargé d'affaires, shifted, it would require major repairs. The Royal Chapel, the King's jewel, was damaged, and as for Malia 'Imakulata, the Catholic church where Mary had had a vision of Deb's Tongan killer, it had about shattered. Coral blocks twisted and separated. A white statue of the Virgin tumbled from her perch on the church's peak and did a somersault, drove herself feet first into the ground.

And Yellow Pier broke apart.

The causeway the American soldiers had built in 1942, the place Deb loved to swim, where she'd walked out from the party at the Dateline five days before her death and gotten sick, where Emile had held her and called her a princess, where it was said that Dennis got angry at Deb for taking off her dress. Its footings shivered and cracked, its plates buckled, swayed, and spalled. Out at the end where the volunteers went swimming, stairway, ramp, and platform sheared and fell into the sea.

29

American Taboo

One by one, officials involved in the case refused to talk about it.

Mary dealt with me through a relative. It had been about the worst year of her life. She didn't see why a reporter should be interested now, reporters weren't interested back then. And though she gave some thought to helping me, she said she didn't want to get herself in another traumatic state by recalling the episode.

The men at headquarters who had managed the matter were John Dellenback, Jack Andrews, and Russ Lynch. Lynch was dead. The other two men wanted little to do with this book.

Dellenback demurred politely at first. His memory just wasn't sharp about events so long ago. "What if I were to present you with a factual narrative," I said, "supported by documents and people who were in a position to know, and that narrative showed that this was an important case, and the killer walked away— would that be cause for you to say, We screwed up?"

"I don't have any idea. I would say that if you present that to me, I do promise you that I'll take a look at it."

But a year later I called and was ready to do that, and Dellenback declined. He didn't think it was good for Peace Corps, and he was old, he didn't see the point of making time for this. I urged him to help me because the matter haunted people, but Dellenback, the former president of the Council for Christian Colleges and Universities, said he didn't trust this project to bring healing or closure to anyone.

"I'm afraid I can't be of much help," Jack Andrews said on the phone. He kept pointing out that Dennis was found not guilty. "What was the court decision?" he said.

"Not guilty by reason of insanity," I said.

"OK. Not guilty by reason of insanity. But not guilty—"

I asked him whether Dennis had killed Deb, and he said, "I have no idea whether he did or not." Jack, who had sat through the preliminary hearing in the case, heard the witnesses from Ngele'ia, met with Clive at a time when Clive was telling everyone that Dennis remembered a struggle at Deb's house.

Not much about the case could be made public, Jack said.

"Why not?"

"It could be embarrassing to both governments."

I asked Jack about Robert Flanegin, the chargé d'affaires, and the psychiatrist the American government had hired.

"I'm not familiar with either Mr. Flanegin or the psychiatrist."

"His name was Kosta Stojanovich."

"I never met the man, nor did I ever hear of him."

But Flanegin's cables to the Ambassador said that Jack Andrews had called him from the airport in Fiji to discuss the case, and Jack had overlapped with Stojanovich by a day in Tonga, and had surely met him. His own cables to Dellenback talked about the doctor.

PSYCHIATRIST HAS BEEN INVALUABLE AND HAS WORKED
EXTREMELY HARD DAY AND NIGHT SINCE ARRIVAL . . .

Jack was shutting me out. I forced an uncomfortable meeting with him at his home in southern Maryland, and he wondered who could be interested in this, and then his face flushed when I said that I'd finally gotten a publisher. He didn't see why he should be in the book.

"You held a political appointment, you're in the book."

Jack's former aide who had dealt with Tonga, David Ingram, a former volunteer, also declined to help me. And a couple of former government officials who had participated in Dennis's release said that they were afraid to talk about it for reasons of personal safety. "He's still out there," one noted.

Other former officials did come forward.

Jack's former deputy, Frank Guzzetta, now the chief executive at Hecht's department stores, said that he and Jack had discussed the fact that Mary was a "terrible underperformer"—this as her salary went from $28,500 to $35,000, nearly a 25 percent hike, over her year and a half of abbreviated service.

I left a vague message with Paul Magid about Peace Corps in the 1970s, and when I reached him later, he said, "I thought you might be calling about that case." He helped me out. So did Dottie Rayburn. She worked at the State Department now. She ignored my calls and letters for years, then one day I got her on the phone and she seemed to need to talk.

"My great regret is not forcing the issue. The young man when I first met him seemed unglued to me. It was my first trip to the field, and he was standing in the Peace Corps office there, and he was disheveled and dirty. I said, This is not going native, this is not right, this is not Peace Corps. I told Dick Cahoon, but Cahoon was a psychologist, and I think Dick felt he knew how to handle him.

"I just felt he should not be there. I have had that burden since then."

But as for Cahoon, I managed only a brief meeting with him, and when I asked about Dennis, he looked at the ceiling and said he didn't remember him. Didn't remember the biggest psych case in Peace Corps history. Didn't remember flying to Tonga to visit a killer in his cell. I wondered if Cahoon looked on his relationship with Dennis as a professional one, and therefore private. But at the time he was holding a presidential appointment.

Cahoon's former associate director in Samoa, Brad Dude, helped me, feeling that the case had been hushed up, as did Dick Hailer, Jack Andrews's deputy.

"It was a major event and didn't get to the light of day," Hailer said. "I could see the headline. Danger in Paradise."

Of course there was no such headline. The news didn't even reach the top two former staffers at the House Committee on International Relations at the time, the men who dealt every day with Peace Corps, who had toured several Peace Corps operations to prepare reports on volunteer quality. They were stunned to hear it from me.

"That could have been a huge issue," John Chapman Chester said. "We would have had to look into it and get a full report. It seems to have been hushed up."

"What! Holy gamoley!" John Sullivan said. "I'm astounded that I never heard about it. We should have heard of it."

A couple of former Peace Corps officials tried to suggest that the State Department botched this one. For their part, State Department officials were less than helpful. Martha Sardinas had traveled from Suva out to Tonga during the trial. She still worked for State, she declined to discuss the matter.

Then there was the chargé d'affaires.

In November 2000 he brushed me off with some neat turns of phrase: "This appalling event, arising from the apparent derangement of another Volunteer, still stands out after twenty-five years. After some rumination, I must conclude that my aging memory cannot reliably take me past the outlines of the tragedy, with which you are already familiar from your research."

I wrote back that I could not accept his refusal, the matter was too important. Then I wrote and called again and again, with no response. Once I got Flanegin's wife on the phone, and she said that it wasn't actually his aging memory, he had a strict policy against discussing official duties.

"The Tongans wanted this case to go away!" she said.

I said I wanted to hear that from him, not a surrogate, and then wrote to say, Even if the Tongans did want it to go away, and a lot of them hadn't, didn't the United States wield any power in the situation? After that, Wayne Gardner got in touch with me, and in 2002 I sent Wayne's letter to the chargé, and when I still didn't hear from him, I drove to his place in a retirement community near the Research Triangle in North Carolina.

Mrs. Flanegin came to the door, she said he was down with the flu. We sat for a half hour on a bench in the courtyard. She had delicate cheekbones and artistic jewelry, perfectly coiffed silver hair, a houndstooth suit showing off slender lines (she'd dressed for lunch with a friend, she explained). She said she and her husband were offended that I'd called her a surrogate. I apologized.

"Bob and I thought the father's letter was fishy," Mrs. Flanegin said. "Where was he? Why didn't he do anything?"

I explained what Wayne had gone through, the divorce, and estrangement from Deb, feelings his own daughter didn't like him, he hadn't been there when it mattered. She promised to put in a word for me with her husband. He ignored the next letter too.

When Wayne heard what Mrs. Flanegin said, he got upset. "Should I have been there with a gun, is that what Flanegin was saying? Waiting for Dennis at the trial? Ask that guy, Where was he?"

Wayne said he had a fantasy of getting all the government officials who wouldn't help me on a scaffold in his backyard. Kissinger, too. When I'd sent Wayne the documents in the case, he had noticed the name KISSINGER at the bottom of every cable going out from Washington. "He signed everything, I hold him responsible."

"That's just the form," I said. "The Secretary of State's name goes at the bottom of cables. RUSK. VANCE. KISSINGER."

"I managed an insurance office in Anchorage, and nothing went out with my name on it that I didn't see."

"Kissinger says he probably didn't even hear about it."

I'd sent Wayne Kissinger's letter. "Not surprisingly, I have no present recollection of the matter," Kissinger said. "Your note indicates that I signed a letter of condolence to Ms. Gardner's family, I am glad to acknowledge that fact. It is improbable that my involvement went beyond that. The murder and the prosecution of the crime in Tonga does not appear to have presented an issue of substantive foreign policy. . . ."

Wayne wasn't convinced. "I don't believe he doesn't remember. This is an odd-ball incident. There have been plenty of deaths, but no murder by another Peace Corps worker in the whole history. Kissinger doesn't remember." Then he laughed.

Of course the main person on the backyard scaffold would be Dennis. Wayne wanted justice. That meant putting Dennis in prison, and if the government failed to do that, well—Wayne had old-fashioned Western ideas of what should happen to Dennis. Wayne even showed me his backroom armory, a collection of military knives something like the Seahorse, knives designed for killing a human being.

"What if Dennis apologized to you and sought your forgiveness?"

Wayne shook his head. "Too late."

"What if the government apologized to you?"

"Nice, but not enough."

"What if people helped build a hospital in Tonga in your daughter's name so that her memory will live?"

"Very nice. It's not justice. Justice means depriving Dennis of liberty."

"People say, But it's been twenty-six years."

"What does twenty-six years matter? The truth just came out two years ago. I wasn't angry when I thought justice had been served. A lie lasts forever."

The next time I saw Wayne was late fall, and he said Craig was coming to the cabin. He'd borrowed Wayne's truck to go hunting in Oregon and would be coming in late that night. Craig wasn't keen on what I was doing, but at least this way I could explain it to him. We'd all go hunting together the next day.

It was a Thursday night. We waited for Craig all evening, through venison stew, then at ten I went to bed in the upstairs bedroom. A few minutes later I heard Craig come in. He talked to his dad about hunting. He'd gotten a bunch of chukars, he sounded excited.

We met in the morning by Wayne's truck. Craig's golden retriever, Deacon, was in the bed, Craig was standing at the door. He was about five-foot-eleven, wiry and muscular, with long dark hair framing a narrow face. He was a thoroughgoing environmentalist. He wore old hiking boots and he had a shotgun he'd kept lovingly since he was a young man.

All day we hunted in a private reserve in the mountains outside Peck. Craig could shoot, he depended on hunting for food. He soon had several pheasants, storing them in his vest, and he wore me out hiking up the frozen draws and slopes and ravines. I didn't get a thing, though Craig tried to guide me. "You take

Deke, and just watch when he gets birdy, he'll get real focused and go back and forth, then follow him." Wayne hung back in his own patient hunting style, and to give us time to talk things over. I didn't bring my project up all day long.

Then when we got back to the cabin, Craig had to deal with a wildlife emergency in Alaska by telephone. Caribou were massing along the Steese Highway, making them easy targets for hunters. Craig issued an emergency order, closing down the hunting season weeks before it was supposed to end. We ate leftover venison and drank beer and scotch and then it was nine at night and Craig was flying back to Alaska in the morning, and I realized I was going to miss my chance to say something.

I was sitting on a step in Wayne's living room, when I asked Craig if there was anything I could tell him about what I was doing.

Craig gave me an angry look.

"I do have one question. There are so many things that are wrong in the world. Why this? Why did you feel that you should write a book about something for which my family never sought any attention?"

I'd never thought about it that way, still I tried to answer him. Yes, it was an opportunity for a journalist when a man killed someone and then walked away and the story never came out, I couldn't deny that. But I hadn't really done anything about it for years, till I'd seen his sister's picture and met his mom in 2000. I liked Deborah Ann Gardner at once, and when her mother said that Deb had requested a transfer because she was afraid of Dennis, it burned me up.

We talked for a couple of hours. Wayne watched us silently most of the time, now and then he butted in.

"Tell him what that judge said about Debbie's character," he said. "Tell him the things the government blacked out when they gave you documents. . . . Tell him about the Tongan prosecutor who said the Peace Corps had no pity for Debbie."

"When Craig wants to hear any of that, I'll tell him everything I know," I said.

Craig said he could see where what I was doing was important for his father, maybe good for his father. But Craig didn't want to open it up because he had come to terms with it more, and didn't want to get mad again. A long time ago he'd been very mad and very sad but there was nothing he could do with that, so he'd tried to forge an acceptance with the fact and live his life. He never wanted to think about the guy in Brooklyn. "I'm not vindictive, I just don't want to think about him."

This was the most important event of his family's life. That was true for his father, for himself, and for his mother. It had affected each of them more strongly than anything else, and determined the shape of their lives in some measure. He had lost his sister. That was the only thing, he said. He had lost his sister. Then it was evident that Craig spoke from no principle but from the deepest hurt. "You can put in your book, She had a brother."

30

Condolence Calls

Look at your hair, it looks like it's frosted," said Alice's mother.

Her lovely chestnut hair had gone salt and pepper, and another saleslady at the Bon said, "Whoever did your hair, Alice?" No one had done it, it had just happened. Deb had been dead eight months now, and to one another, Alice's friends said her grieving period was a long one, but Alice was just trying not to think of anything. Because if you blocked it out it couldn't hurt you.

At the other end of Garden Court from her condo lived a doctor who worked at Western State Hospital, a big mental hospital a mile away. Alice thought about knocking on her door, or saying "Excuse me" when she saw her in the cul-de-sac. What happened to a man like Dennis? When did he get out? What was to hold him?

She wondered if the doctor at Western State could somehow get hospital records. But Alice didn't ask. You could feel stupid asking about things no one knew anything about. Besides, what if she found out Dennis wasn't going to go away for a long time? That would just make her angry.

One day in June, the phone rang.

"Mrs. Gardner, I hope you remember me, my name is Frank Bevacqua."

Her wail cut Frank off. "Frank!"

Of course she remembered him. She'd met Frank at the airport not a year before, to give him celery salt and *Vogue* magazines, and then two months later Frank had written a letter. "I just want to offer my heart to you. I loved her immensely, maybe differently from you, as her mother, did, but she meant the world to me. . . . If there is anything at all I could ever do for you, to help you in any way, please don't be afraid to ask."

Well, Alice asked.

"What happened, Frank? Do you know what happened to Dennis?"

Suddenly Frank didn't know how to help. He'd heard the story about Dennis, but he didn't know for sure.

"What did Peace Corps tell you?"

"They didn't tell me anything." She had read in the *News Tribune* that Dennis was going into a hospital in Washington, D.C., but the Peace Corps had told her nothing, and Alice didn't know who to call.

Then she was crying, and it was the most uncomfortable phone call Frank had ever made. She was so distraught, he didn't know what to say.

"I don't know, he was going back to be hospitalized." Then Frank added feebly, "I'm looking into it," and hung up the phone, shaking. He'd never experienced such hurt in another person. What is going on? he thought. What is the Peace Corps doing? The Peace Corps just dropped her like a lead weight.

Alice sent Wayne the clipping, PRIVEN BEING SENT TO HOSPITAL, and copies of various letters she'd received, from relatives, from girls at Tonga High who had set up a small scholarship fund in Deb's honor, from Frank Bevacqua and Mike Braisted, and from public officials, too.

I have learned with deep regret have just learned Mrs. Rockefeller and I want
to offer our deepest truly distinguished herself I hope you will find some
measure of placing service above self It is my hope that you will find some solace
in the gave her life the cause of service to others. Please accept Sincerely
Sincerely

Wayne glanced through the letters from the vice president and secretary of state, but they just made him angry. The signatures were real, but the word "service" was bogus as could be. Debbie hadn't died in service. She was killed by another volunteer. Her death hadn't helped the United States one iota. Not that he issued a peep of complaint. The letters were form letters, what did Nelson Rockefeller or Henry Kissinger care about his daughter?

Wayne didn't want to think about Tonga. He drank hard and did his work, but he hated every day he had to wear a tie, and when a salesman said he was sure sorry on the phone, Wayne got right off, "Thank you" and click. He wasn't going to talk to anyone about Deb.

There was one time he did. A man he worked with was a hurt father. The father was born-again, he had strict rules for his children, and when his daughter turned 17 she came home after curfew and the father flipped his lid.

"At eighteen I can do what I want."

"You're seventeen and you're living in this house," the father said, "and as long as you're living here you'll be under my rules."

So his daughter moved out to a girlfriend's house and started doing what she wanted. But a man still had power over his daughter. He could cancel her driver's license, he could freeze her bank card, he could call up the people she was living with and say, "She's your responsibility."

Wayne asked the father to come into his office and close the door and tell him what was happening. A father could get a divorce from his own daughter, the man explained.

Then Wayne told him about cutting off his daughter.

"My daughter died before I got a chance to make up with her. You are a hurt father. I want you to go back and tell your daughter you love her and she'll always have a home.

"You are her father and her mother, and you love her very much. You are not going to take away her rights. I never want to see another father go through what I'm going through."

The father's eyes got big and he said nothing and got up and walked out and that was one time Wayne thought maybe Debbie's death had a reason. The next day the hurt father came back to Wayne's office, and he was in tears. He thanked Wayne and shook his hand, then a week later his daughter came back home, and the father went back into Wayne's office.

"Wayne, I am praying for you," he said.

"Somebody should," Wayne said, and walked out on him.

He'd stopped praying. The former member of the board at the Lutheran church in Chicago didn't believe in churches, and he was angry at God. He didn't stop believing in God, he believed in God more than ever. Someone had to be responsible for Deb's suffering, and when Wayne died he would go after God.

You sonofabitch, when I get up there I'm going to kick your ass, I'm going to castrate you, Wayne said to God.

A Christian friend told Wayne he had to have faith, God worked in mysterious ways. After the apple in the garden of Eden, God had given man the freedom to make his own decisions. It was a bad man who had done this, and in the end God would take care of the problem. . . .

So after that Wayne's mind turned back on himself, on the choices that he'd made with free will.

The cold war wasn't Debbie's responsibility. What did a child know? A child didn't love her parents the way a parent loved a child. Parents were responsible for the child's being. A child would love and respect her parents, but the Bible said you leave your mother's apron strings and tie up with your spouse, you go out on your own, just the way Debs had gone into the Peace Corps. But to a parent, when his child left, half his insides were torn out. He remembered how he felt when he heard that Debbie had gone to Tonga. He hadn't even known what Tonga was, and a fear had possessed him even then, that he wasn't going to be there to protect her, and what would happen to her so far from home.

A father was responsible, no matter how old the child.

When Wayne had a drink he sometimes thought about the time he'd actually been Debbie's father. He'd been there when she tripped and fell the first time, been there when she tumbled off a bicycle, been there at every major good and bad in her life, then what—he was at the Arctic circle under the northern lights, and Debs was dying in Tonga.

Guilt clenched his spirit. He'd deserted his daughter long ago, he could never be forgiven for that.

He wondered about the boy's parents, why they'd never gotten in touch with him. If it had been his son he would have called the other parents. It would have been a hell of a hard call to make but he'd have done it. To express remorse and also to tell them what he'd done to see that his son had been put away.

Barbara's mother was religious, a farm woman from eastern Washington. She came to visit the Gardners at their new house near Flattop Mountain in Anchorage and she started to say grace at the dinner table and Wayne got up and left the table. "No one's going to say grace in my house."

Barbara's mother kept trying. The next night Wayne asked for the mashed potatoes, and she said, "You'll have to say a prayer before I give them to you."

"All right, fuck God and pass the mashed potatoes."

Barbara's mother sobbed with Barbara in the recreation room. Wayne Gardner had a hard heart, Barbara saw the softness only when he had a lot to drink. Then he would start to cry and say, I'm trying to forget.

One night when Barbara was going in to the night shift at Western Airlines she met Wayne in the road coming home from a bar after work. They stopped their cars alongside, and Barbara got out to say hello and said, "Wayne, what's wrong?" and Wayne broke down.

She stood out there in the road, trying to hold her husband through the truck window.

"I know that it hurts, I just don't know how to help—"

Later they were in the house and Wayne said, "I'm going to take my gun to New York when that guy gets out of jail, I am going to pay Dennis a visit."

"Wayne, what is that going to prove? You'll be the one going to jail for shooting him. It's not something Debbie would want."

Wayne's first marriage lasted sixteen years, this one lasted sixteen too. After this divorce, though, they stayed in touch, and from time to time Wayne asked Barbara spiritual questions.

"Barbara, tell me something, do animals have souls?" he said. "And something else I've written down to ask you, where was Jesus between ages twelve and thirty-two?"

Barbara talked to her minister. No one is sure about Jesus. The Bible says he was out on public missions. He was probably in his home town of Nazareth, helping his father the carpenter as carpenters' sons do. We never read anything about Joseph again, and some people say Jesus' father died and Jesus took over the business, but they're guessing. Animals don't have souls, the minister said. They feel guilt when they've done something wrong, but it's because they know they're going to be punished. It's the punishment they're afraid of, they don't really regret what they've done.

"Do you ever pray, Wayne?" Barbara said.

"I pray but it's in the mountains when I'm by myself. I don't ever pray for myself, but for others."

"Well, that's great, why don't you try asking for something for yourself?"

"I'm a lost cause."

"We've all done bad things."

"I know mine. Every day of my life, they're right in front of me."

"But why don't you just once pray for yourself?"

Wayne shook his head.

31

The End of the Legend

After Frank Bevacqua talked to Alice on the phone he went to Washington. He still carried Deb's photograph in his wallet and he wasn't going to let Deb's memory go without some kind of fight. He went to the offices of that region that was no region, NANEAP, and Dottie sent him to the counsel's office, and Paul Magid took Frank into a conference room, closed the door.

Turned off the light, Frank says. Don't remember that, says Magid.

"This conversation never took place," Magid announced. "This is the first and last time I am going to talk about this to you. We brought him back under an agreement that he would be hospitalized. He refused to be hospitalized. There was nothing we could do. A psychiatrist said he had a situational psychosis, and he couldn't commit him. So he went back to New York.

"And you cannot tell anyone what I have told you because it is a violation of privacy law. If you say anything publicly Dennis can sue you."

Frank was even more shaken up than he'd been after talking to Alice. He thought about saying something, but what could he do? He was one guy, still living in the South Pacific, he wasn't going to be able to fight the government or Dennis.

He thought that Dennis had played everyone like a card game. He had said nothing publicly from the beginning to the end, meanwhile counting everyone else's cards. Frank thought of how Dennis would say with a laugh at the end of a poker game, "Why did you take the jack of diamonds when you had the ten?" Frank had been a witness for the Crown. He wondered if he was in danger.

There's nothing I can do about this as a person alone, Frank said to himself. He'd been told point-blank, Whatever you think happened, nothing happened in this country, Dennis was never judged in this country. He thought of calling Alice, but he thought, Alice is trying to heal, there's no point in throwing horrible facts at her.

He was just going to have to forget this. Then Frank thought, how could he forget Deb Gardner, he couldn't. He kept her picture in his wallet for years till finally he understood it had too much of a hold on him, he had to put it in a box somewhere. Snoopy wasn't going to do any backtracking, not in this direction at least.

Vic Casale got the news in Denver. Vickie called him from the Blue Mountains and cried in disbelief at the turn of events. Dennis had even tried to call her when he got back, as if things could still go on as if nothing had happened, and when Vic went home to Lindenhurst he borrowed his dad's Dodge and drove to Brooklyn to look for Dennis. The Brooklyn book listed a bunch of Privens. Vic sat outside one house then another for a while, waiting and thinking.

He thought about a time in Tonga. He and Deb were at a party at Sela's, A3Z was on, and Gideon the dj said something about white man's blues, that the white man gets the blues too, not just the black man, and he put the needle down on so, so familiar organ chords and the volunteers cried out at the acidy anthemic dirge and Deb and Vic held each other like the high school prom—

That her face at first just ghostly
Turned a whiter shade of pale

before Gideon said, "Oh you'd never think something so bad could happen to a white man he'd sing about it. That a white man could put such a hurt on another white man, but it does happen, it does happen—that kind of music gives me goose bumps all over, and I mean *all* over—" And he laughed at his own foolishness, and Deb and Vic laughed at him too.

Maybe I'll talk to him, maybe I'll kill him, Vic thought. I'll just wait and see what he does.

But a day of that only made him feel more rotten and useless so he stuck to Long Island, and a couple of days after at a mall near his parents' he opened the car and his bowie knife was gone. The knife he'd taken everywhere for six years, the knife he'd held to the throat of the Tongan guy who'd had ideas about him and Deb. Now some prick in a mall had lifted it.

Vic slammed the car door with disgust.

The McMaths learned the outcome of the case in Perth. Sidney Priven sent a letter to John McMath, thanking him for the friendship he'd offered his son in jail, and Sidney said Dennis was back home, planning a trip into the mountains to do some thinking.

John reasoned that the father was doing what he could to tie up broken ends. Sidney had been denied the satisfaction of telling friends and relatives, Dennis was in the Peace Corps and he did thus and such. Saying thanks to someone who'd shown kindness to his son, a surrogate father in Tonga—that gave Sidney a place to stand amid the shame and hurt.

Marie's response was more emotional. She wondered if Dennis's father and mother were afraid of him, if Dennis had stood over his father telling him what to say. When John wrote back, Marie put a letter inside addressed to Dennis with "Marie" at the bottom and with no words of her own, just a Christian song of praise.

Something beautiful, something good.
All my confusion He understood.
All I had to offer Him was brokenness and strife
But He made something beautiful of my life.

Marie wanted to forgive Dennis, to rid herself of the feelings she had borne since that afternoon because they were tearing her up. Dennis had destroyed a whole family's trust in him, and she couldn't even tell her children or her parents what had happened. She prayed that God would use the words to touch Dennis and change things. Because Dennis was a mess, and he was a mess that had sealed itself over. Marie knew that he had the ability to put something in a box and put it away in his mind. Dennis never talked about anything he didn't want to talk about, didn't think about anything he didn't want to think about.

But Dennis didn't respond, didn't apologize.

There's a point where the spiritual and the psychological end and the political begins. Knowledge of Dennis's freedom ate at Marie. She remembered the Peace Corps closing ranks after the murder. No one in the expatriate community knew just what was happening, and they were never going to find out, and Mary George was going around saying Dennis hadn't done it. "We're all supposed to walk around with bags over our heads," Marie had said to John. "Yes, well it's not adding another stripe to the American flag, now is it?" he responded.

Marie wondered what the Peace Corps had told the girl's parents, and she wanted to tell them herself. If she only had Deb Gardner's parents' address she would write them and tell them what Dennis had done to her, and tell them that now Dennis was going into the mountains to think.

Emile was home in San Bruno, trying to put his life together. He stayed with his folks and lived off his readjustment allowance and thought about Tonga too much. He went out walking at all hours, and his mother whispered to friends, His girlfriend was murdered.

Emile got a job at the nearby Sears doing window displays, and one day he came home for lunch and the front door was open, and there was no sign of the dog or his parents.

The door was never open. His parents didn't leave it open.

He stood outside the door for a long time, saying, Hello, hello, hello, hello, louder and softer, just like some Tongan kid back in Ngele'ia. Then the neighbors drove up with Emile's parents and Charlie the dog. His parents had gone visiting and forgotten to close the door, and Emile understood, this will never be over.

Lampshade Phil had become a reporter back in New Zealand, Emile wrote a letter to Lampshade.

"How they hang'n? It's your old buddy 'Emili, remember?

"A tat strange about Dennis, don't you think. Boy am I glad that shit is over. The word's out that when he got back to the States he was set free. Great!!!! That's why the USA is the way it is today. But enough of this bullshit . . .

"All the *fokisis* went to Fiji for R and R around October and hadn't returned when I left in December but I didn't care . . . Don't let your meat loaf, 'Emili."

Lampshade Phil worked for a little paper in Thames, but it wasn't like he could do anything with the story, it wasn't a New Zealand story. It was inevitable that American writers also heard about the story and tried to do something, but they were too far away from the Kingdom, and so the story stayed a legend.

A freelance writer in New York called Charles Salzberg, who had read about the murder in the *New York Post*, mailed questions to the Peace Corps, but the letter he got back from the associate counsel did not give him much to go on and he soon dropped it. Then Joe Theroux, a Peace Corps volunteer in Samoa who had heard about the case, pressed Mike Basile to tell him about it when Mike passed through Samoa, and Mike sat with him for a couple of hours in the bar at Aggie Grey's. Theroux was inspired to write a long story, "Island Fever," about a cipher called Popo Bob who was accused of the stabbing death of a sweet uncomplicated American girl. Theroux focused on the paintings the man made on his house: "[A]round the archways of the door were circles with lines through them, triangles dominated by X's, misshapen rectangles, again with those small circles, all indecipherable hieroglyphics, with a meaning locked away in the twisted mind of Popo Bob." But Theroux was not able to publish the story, and Mike Basile was disappointed.

Then the Internet came along, and the murder got onto the Internet.

An Oregon minister put a sermon about the case on-line, after hearing about it from Pila (Bill) Mateialona, the former Peace Corps driver.

"Am I going to die, Bill?"

"Hang on."

"Will you send me home if I do?"

"Of course we will. I'll make sure."

In the sermon, the Carter Administration stepped in to bring the killer home, where he was institutionalized for a number of years.

Another story on the Net had a *pālangi* called Andy telling tourists at a Nuku'alofa bar about the murder. The twist was that Andy was the killer. He'd gotten out of a mental institution and come back to Tonga.

"She was a beautiful girl. Cassia. It means 'cinnamon bark.' . . . She had a . . . I call it a china doll's face, long and oval and very white. Her hair was pale red, the color of the clouds at sunset."

Peace Corps headquarters moved, and if anyone talked about it, they called it the Tonga murder, and didn't know very much about it. The documents in the case went into boxes and got shipped to the National Archives, and meantime those with a real connection to the events began to die.

The lawyer Russ Lynch who said, "Once out, all out," died, and nothing was out.

Judge Hill who said she was of loose moral character and Dennis was unlucky to be involved with her died in Hammersmith, London, a lonely end.

Ambassador Selden died, and so did his wife Mary Jane.

The Peace Corps director John Dellenback died. Dr. Rosa, the chief medical officer at headquarters who had taken Dennis to Sibley, died. Dr. Lebensohn, who saw Dennis there, died. The former head of the State Department's consular desk who directed the chargé d'affaires not to take custody died.

Warren Fortier who went moose hunting with Wayne died. Grandma Agnes who said she wanted to be a bird of knowledge and fly out and bring back some knowledge of every nation and country died, and so did the other three grandparents.

Lolo who fought with Mary over writing about the murder died. David Mclean who had vacationed with Deb in Ha'apai, died.

Ruth Judge who'd fasted and done deep prayer for Dennis died.

Tongans were dying, too.

Vaka Pasa who drove the truck to the hospital died. Puloka who tried to save

Deb at Vaiola died. Faka'ilo who interrogated Dennis died. Maeakafa the superintendent Mary believed had killed Deb died. Dennis's bush lawyer Tomi Finau died.

The members of Privy Council that had voted to release Dennis dropped one by one: Tu'ipelehake, Tuita, Ve'ehala, Tupou.

The Police Minister died.

'Akau'ola felt death coming. He needed help standing up, he had a trusted man walk alongside him at official functions and pretend that he wasn't holding him. Diabetes was stopping the circulation to his legs, still he taunted his worried brothers by eating chocolate in front of them and swore by his Christian folkhealer in the bush and disdained Western medicine even as he quoted Shakespeare in the newspaper.

One night the minister of health had a dream that 'Akau'ola had approached him and then 'Akau'ola was on a stretcher, and the doctor went to two of 'Akau'ola's friends and told them about the dream and said he could cure 'Akau'ola. Still, 'Akau'ola refused treatment.

He went on vacation to his chiefly lands in Vava'u and collapsed while he was fishing and was brought back home. Now at last the doctors saw him and they could not help him, he had gangrene, and when his officers went to visit him in Vaiola Hospital, they said 'Akau'ola's eyes roved the corners of the room, he was in spiritual pain, facing the men on the scaffold he'd put to death.

A medical team came out from Auckland to take him to a hospital in New Zealand, but on the way to the airport 'Akau'ola passed out again.

So it was decided there on the road, There's no point in taking him to New Zealand, he'll die on the plane.

How often 'Akau'ola had chattered to his adopted daughters about living out his old age in New Zealand. But that was another romance the minister told, New Zealand held no appeal to 'Akau'ola, not even with red-haired June of yesteryear. His place was Tonga.

They turned the ambulance around and at four o'clock that afternoon in Vaiola, 'Akau'ola passed. It was an 'eiki death, a chiefly death: He had smoked cigars and drunk coffee to the end, he had met his death standing up.

That was the last day of 1995. A year later Michael Basile felt he had to do something and began writing letters to his Peace Corps buddy, and Jan Worth started a novel about 1970s sexuality and, still angered by the murder, put her ad in the newsletter, and I met her and Mike and together we blew the legend to kingdom come.

32

Behind the Bricks

In Nuku'alofa I always dropped in on the former Peace Corps office manager Makaleta Fifita. Maki was busy and tough but she cut me a break because my work had reconnected her to a lot of people she loved. She would never forget Deb sharing the Sears catalog with her, or Dennis following Deb into the mailroom. She wrinkled up her nose at that memory. *"Namu peka."* Smells like a bat. But for Maki, Dennis was still in the human family, and she shook her head over Wayne. The only way Wayne Gardner could clear his life of anger and vengefulness was with Dennis's help.

Maki said that Dennis had to do what Tongans had always pressed him to do, apologize and express remorse. That way God would give Dennis new life and lift the cloud from Wayne. The way I understood Maki's words was that Dennis's disbelief in God, or Deb Gardner's soul, or her humanity, however you wished to phrase it, had been communicated to Wayne. Dennis continued to exert a dark power over Wayne.

It was time for me to go to Sheepshead Bay.

Dennis lived in a brick complex at the edge of Brooklyn. Six stories, a small courtyard, a man sitting on a bench with a Holocaust tattoo on his arm. Dennis's door was at the end of a little darkened hallway. You came to a point where it was the only door in front of you, like a doorway you had to go through in a dream scene in a movie, the dull co-op walls slanting toward one another and the dark red door in front. From that perspective, it looked like the lid on a box.

I'd put off knocking on that door, and thereby taken the risk that Dennis would learn about my project secondhand. Some people seemed to feel continued loyalty to him. One former Peace Corps official had hugged his rib cage defensively as we spoke and urged me not to mention Dennis by name. "You could wreck his life," he said. I wondered if Cahoon, who had visited Dennis in jail and said that anyone who was haunted by these events should be talking to a therapist, didn't feel loyalty to him too.

On the other hand, several former volunteers offered to ride shotgun with me to Sheepshead Bay. A couple were still angry at Dennis for having damaged the relationship between the Kingdom and the United States, for disrupting their work, and wanted to tell him that. Robert Forbes, the volunteer who saw this project as providential, said he was willing to accompany me and drafted a letter that made a spiritual appeal to Dennis. "Speaking your feelings now about what transpired twenty-six years ago might help bring some closure to an issue that has bothered many people since then," he wrote. "I also think that what you have to say about the murder and how it has affected your life since then will strongly influence how the book's readers judge you."

But Dennis didn't know Bob, and even if his company made me feel more comfortable, I didn't think it would help. One volunteer who might make a difference was the Bird Lady, but the Bird Lady still had some *"Faka'ofa a Tenisi"* feelings and I didn't feel that way. And though Emile had come out to Tonga while I was there, a couple of times, to give equipment to the fire company and start a scholarship in Deb's name, the idea of inviting him to Brooklyn was just too fraught to try and push with him.

By then Emile had told me about his last visit to Dennis in jail, a few weeks before the trial.

"I have an idea," Dennis had said with the air of someone who had arrived at a solution to a mathematical problem that had plagued mankind. "It's simple. They hold the trial. At the end they find me guilty. Then you come forward and confess, you were the *pālangi* with the beard the neighbors saw. You killed Debbie. The police will have to let me go and try you instead.

"But then when you're convicted, I will come forward and say that I did it after all. And then, see—they can't charge me again because of double jeopardy? In the end both of us go free."

Emile had looked away. "I don't know about that one." And after that he never went back.

I felt that I needed guidance before I actually knocked on Dennis's door, and I turned to Kevin McKeown, an investigator and author (of the book *Your Secrets Are My Business*), who another writer friend had hooked me up with. In August 2002, Kevin agreed to coach me about the approach.

"OK. Sunday, twelve o'clock, Yonkers train station. I'll be driving a white station wagon with New York plates," Kevin said.

Just before I took off he called me back.

"Change in plan. I'll be in a maroon van with tinted windows. D.C. plates."

Kevin was good-looking, compact, charming. He was with a friend, a retired New York detective. Charles Meleski was calmer, gray, dignified, splay-footed. They took me along on a couple of jobs that day: a drifter in the Bronx whose body the city had somehow lost, a marriage case across the river. They were trying to help the drifter's mother locate her son's body, but the tinted windows were for the marriage job, so Charlie could set up in the back of the van with a camera.

They were as street-smart as anybody I'd ever met. They seemed to be able to go wherever they wanted by flipping out their wallets and telling a two-line story. At a police precinct in the Bronx, they strolled into the detectives' room and I followed. "Did you see the monkey in there?" Charlie asked when we left. He meant some guy behind bars next to the detectives' room. I hadn't noticed.

We grabbed a late lunch at a diner in Manhattan and I showed them all my pictures from Tonga. I got out the lineup shot from Ngele'ia and Charlie touched Dennis's face.

"That's your guy, right?"

They both pushed me to knock on Dennis's door. They would back me up. Kevin would call Charlie's cell from his cell, I'd slip Charlie's cell in my shirt pocket, and they'd keep Kevin's cell on intercom in the van. That way they could hear everything going on in Dennis's apartment, and if there was trouble they could jump in. Charlie lifted his shirt to show me his service revolver.

But I wasn't ready. I didn't want to knock on Dennis's door with two investigators outside, I wanted to go on my own in good faith. Also, I wanted to have a cellphone of my own so if it came to it, I could give Dennis a mobile number.

We took the Belt Parkway around the edge of Brooklyn and hit traffic at the Verrazano. They started up again about me knocking on the door, and I told them about Marie McMath, what had happened to her that day.

"Whoa," Charlie said. "I wouldn't go in there alone." "No way, that's a sick puppy," Kevin said. I had to laugh.

They told me I should learn as much as I could about Dennis before approaching him. Who he hung out with, what he did, what his routines were. The two of them would get me started. I walked around with them as they talked to some of Dennis's neighbors. We all agreed on a story. Someone we were looking for had disappeared into the building. We were trying to figure out which apartment he might have gone into. The investigators listened sagaciously as people described the various neighbors. Now and then Charlie "tinned" people—flashed his license.

In no time we had information about Dennis. He was an old guy who lived by himself, someone said. He worked for the Social Security Administration.

It was dusk when we came back into the courtyard.

"Last chance," Kevin said. "Go knock on his door. You'll be fine."

But I refused again, and as we got back on to the Belt Parkway, Kevin had the idea of calling Dennis for the heck of it. Putting his phone on intercom, he asked me for Dennis's number.

"NIghtingale-6-3677," I said.

An operator's voice came on. "The number you have dialed has been disconnected. . . . No further information is available. . . ."

"He's on to me," I said. "That number's been in his family for thirty years. You don't just change it."

I felt relieved. The only point in knocking on Dennis's door was to surprise him into unconsidered statements. I'd lost the element of surprise. Now I could do the gentlemanly thing and write.

The next morning, the two investigators' work checked out. Dennis was the area systems coordinator for Social Security's Brooklyn office. That meant he was responsible for all the computer systems in a huge area. The government told me that he had been at Social Security for fourteen years, and made more than $78,000 a year. Perfect. The last guy in the world to accept authority had spent his working life entrenched in the biggest authority around.

A day later I stood outside Social Security offices in downtown Brooklyn to see if I'd recognize Dennis coming out of the office after work. Kevin had given me a primer on tailing someone. You have to overcome your natural instinct against doing this, your feeling that you are being seen. Wear a baseball hat and practice on people. What are they going to do if they think you're following them? You can drop off whenever you like.

Test yourself. You'll get a feel for how far back you can hang and still follow the person.

Montague Street was crowded with people hurrying home. Just about anyone could be Dennis. I followed half a dozen fiftyish chunky guys into the subway and was about to give up when Dennis walked out of the building. I knew him in an instant. The planes of his face were the same strong planes in Emile's photographs. He wore dark glasses and a dark T-shirt, jeans, and sneakers. He was broad-shouldered, balding, had a gray bristling beard and an informal, completely-on-his-own manner. I tried to think what he reminded me of and then I realized, he looked just like a former Peace Corps volunteer.

I followed him into the subway, lost him in the crowd.

The impression of him worked at me that night. Dennis seemed little, unpretentious, unusual. He didn't mind being deeply odd in Brooklyn any more than he'd minded being unpopular in Tonga. Kevin hadn't been able to find a car listed for Dennis; Dennis didn't drive. He had built some strange sort of life for himself in Brooklyn between the co-op apartment that had belonged to his parents and the government office. Vic Casale liked to say that Dennis was "behind the bricks in Brooklyn." And Vic was right, Dennis seemed entombed in a type of institution of his own making. For the first time I felt what Greg Brombach felt looking into the jail cell two days after the murder. Pity.

I sent Dennis a letter. "I'm writing a book about Tonga and the Peace Corps and the events of 1976. I've known about the case since the late 1970s when I went backpacking through Samoa.... I come to you with an open heart, I mean to demonstrate that to you. Will you meet with me to talk it over?" I mailed the letter from Grand Central Post Office, where I now had a PO box, the Friday before Labor Day.

When he'd come back from Tonga, it had stunned his neighbors and friends that he was free. They'd read about him in the *Daily News*, now here he was. Some of his old friends were disturbed. "You wouldn't believe it, I ran into him on the street in Manhattan, he's just walking around," one said to another. "He doesn't seem to be in mandatory therapy or treatment or observation or anything—"

It made the friends uncomfortable. They thought of the girl's family. Was she a Tongan girl? some asked. They thought something should be done about it, but what was there to do? The government wasn't doing anything. They were 24- and 25-year-old boys. How did a person kill someone somewhere else and just come back and walk the streets? How did that happen, and not a word about it?

If you said, "Hi, Dennis, how are you?" he'd say "OK, Joe" and that was all, pass you by without even looking at you. If you didn't say anything, he didn't say anything back. He resumed playing in the old fraternity brothers' weekly poker game, and no one talked about Tonga. Still it was whispered. One friend said to another, "Isn't it eerie, here we are playing cards with him just as we did before he left, as if nothing happened," and the fraternity brothers joked that you should never have sharp objects around Dennis when he was playing poker. His friends didn't believe that he was insane. He was disturbed, they could believe that. He didn't look you in the eye. He was different, he was brilliant, he was nuts. But insane? No.

There were rumors. Friends reported that Mimi seemed to accept that Dennis had done it. Then the American government had done its best to get him off the island and hush the thing up. Dennis may have been high on something when he did it.

Nothing had changed and everything had changed. He had a big beard and he didn't drive, he rode a bicycle everywhere. He was more withdrawn after he got back, more contained than ever.

Some of the friends had trouble with the idea of what the girl's family knew. They were good kids, middle-class men setting out to be upstanding members of the community. They believed in Jewish ideas of *mensch*-dom, charity, and *rachmones* (which is Yiddish for '*ofa*'), they believed in a just society. A few guys said something to Dennis. Fraternity brothers had gotten together at the old house for a reunion, and one asked Dennis, "Is it true what we heard, what happened in Tonga?"

Dennis was casual, he nodded his head. "Yeah, it happened."

Maybe the encounter was planned, but when other questions came Dennis shrugged.

"It's a long story, if we get together some time I'll tell you about it."

Like they should go out and have a drink with the Mets game going over the bar.

It disturbed the friends that a person could put that in a box, and some never talked to him again. Others accepted it. It was temporary insanity, they said, it's the same old Dennis. Then a year or so after Dennis got back, Mimi died, and neighbors said, She died because of what Dennis did. He got some kind of job. When Reagan broke the air traffic controllers' strike in 1980, Emile Hons thought he saw Dennis on television, standing in front of a control-tower computer. By 1988 he was working for Social Security.

Time transformed the Priven household. His father moved to Florida and married a good friend of Mimi's, Helen. In 1990 or so, Dennis got married, too, to a Hispanic woman, and had a real wedding. They got divorced in 1996. A court clerk in Brooklyn said police records weren't public, but when I showed him Deb's picture and told the story, he disappeared and came back and said that it didn't look as if Dennis had been in trouble with the law.

He lived in his parents' place by himself. His father died. He didn't have a car. He rode his bicycle around Sheepshead Bay with his knapsack, the eternal Peace Corps volunteer.

His brother Jay, the better athlete, was a coach at Boys and Girls High School in Brooklyn. I went to the school and asked to see Jay, and the guard picked up a phone and Jay got on. I said I was writing a book about the Peace Corps, 1976.

"I have nothing to say," he said.

I mustered a commanding tone. "It doesn't work like that, Jay, I'm not going away. Your name is in the National Archives saying there was a cover-up, your name is in my book. I need to meet with you and tell you what I'm doing eye to eye."

"I'll come out."

Jay brought me back to an office with a steel bench in it. He was tall and thin, fit, with glasses and a narrower face than his brother's but with Dennis's dark eyes and heavy jaw. He wore jeans and a gray polo shirt with the high school logo on it. He crossed his arms and kept shaking his head and saying, "I don't know anything about this."

Then he said, "I never talked to my parents about it, and what they know they took to their graves. My mother died in 1978, my father died a year and a half ago. I'm the only one left. I don't have a brother."

I showed Jay a photograph of Deb Gardner. "Do you know whether Dennis killed her?"

"I have no idea what happened."

"You told the Peace Corps you thought there was a cover-up."

He shook his head. "Anything I said, I was doing it for my mother, who was sick. I will give you no information."

"The parents of Deborah Gardner wonder why your family never reached out to them to express remorse, why your family didn't see that Dennis was incarcerated."

Jay lifted his hands. "I don't know anything about that. Now you're talking about a family I don't know."

"You don't feel any connection to these events?"

He kept shaking me off now.

"Did you talk to your brother at your dad's funeral?"

"I haven't talked to my brother in twenty years."

I got the same closed-off feeling from Dennis's former wife. She was remarried. Her husband was vacuuming the minivan when I knocked on her open door. The mention of Dennis's name produced anxiety.

"Anything you want to know about Dennis Priven you have to ask Dennis Priven. He does not like Dennis Priven—" she motioned at the man in the minivan.

I told her about the murder in the South Pacific.

She shook her head but didn't seem surprised. "I don't know anything. You have to talk to Dennis Priven about that. Dennis Priven is the only one who knows anything about Dennis Priven."

"Do you want me to give you a letter from my publisher?"

"I just want you to leave me alone."

It rained over Labor Day, and on Tuesday there were still big puddles under the elevated in Sheepshead Bay. I leaned against a stanchion of the Q train stop in the shadow, watching the people getting home from work.

Dennis wasn't going to write me back—Emile said that Dennis would have just two words for me—so my plan was to study his movements and then go up to him on the street. First I wanted to make certain that the guy I'd seen leaving the Social Security office got out of the subway here. I'd follow him back to the co-ops and figure out the best place to have the confrontation.

I wore a baseball cap and maintained surveillance on the bus line. The Russians rode by in their late-model Mercedes sedans, and I tailed a couple of people down Avenue Z to work on my technique. Rush hour was petering out when I turned on my new cellphone, and there was a message.

"Hi. I'm trying to reach the writer of the letter, August 29th—"

His voice sounded sincere, high and sweet, with a strong Brooklyn flavor. He seemed to have a new cellphone of his own. I called him back.

"Dennis?"

"What's up?"

"I want to meet you."

"OK but it has to be short."

"Thursday?" I said.

"You tell me where," Dennis said. He was tongue-tied in the way Emile had described, he almost had a lisp.

"Uptown, downtown, Brooklyn, anywhere's OK with me."

"Tell me where," he said firmly. "I'm like the elephant. If you tell me to sit on the egg, I go sit on it."

We agreed on Dean and DeLuca, the food market on lower Broadway in Manhattan.

"Fruits on the right, meats and stuff at the back," Dennis said. "You know what I look like. How will I know who you are?"

I described myself, and he said, "I'll be wearing a Champion T-shirt, cut off at the arms."

"What color?"

It felt important to maintain an innocent connection on the smallest details.

"Let me look in my drawer," he said. "Here's a nice one. White with blue letters."

33

Getting Away with It

Workmen were eviscerating Broadway at Prince. The jackhammers were going, girls walked by in their slip dresses. From the coffee bar at the front of Dean and DeLuca, I saw a girl whose lace underwear showed above her wrapskirt and thought about how these women would have seemed to the kids in Tonga 16, with their duffel bags and long dresses and nylon underwear. No cotton; it falls apart.

In my knapsack I had documents from New Zealand, the analyses the government had done on Dennis and Deb's clothing twenty-six years before. The Department of Scientific and Industrial Research had shown that the blood that had soaked Deb's dress was the same type as the blood found on Dennis's clothes, under two systems of classification, O and M. It wasn't a precise match, one in five people would test O, M. (The blood had deteriorated too much to be tested for Rh proteins.) All the same, it was science, and Dennis believed in science. I wanted to be ready if he said he was innocent.

Dennis arrived promptly at 12 noon. He was the guy I'd seen in Brooklyn a few days before, wideset, with dark glasses and a fixed lowered oxlike expression. He wore the white Champion shirt and jeans shorts, new white sneakers. We shook hands and Dean and DeLuca suddenly felt tight as a closet. He gave a tilt of his head, and we walked out to the street, he turned east into Prince past the subway entrance.

"Do you go by the standard journalistic ethics?"

"Yes, why."

"So—everything I tell you is going to be off the record."

"OK."

We spent the afternoon together. As we talked, we walked up and down Manhattan, sticking with big north and southbound avenues.

My impression was of a heavy guard and under that an almost helpless tenderness. His charm was intellectual, Dennis's mind was interesting. It could go almost anywhere, and he was funny. I never looked at the traffic as we talked, I was too deep in the conversation, and now and then Dennis reached out and grabbed me, to stop me from walking into a car. The sense that the Bird Lady and others had had that he would do anything for you was now mine. A couple of times he stopped me in a brotherly way to tug the zipper up on my knapsack, an issue he had dealt with on his own with a clever sort of key-ring device that held the two zippers together.

For the first half hour or so, I spoke. I apologized for going behind his back for so long, but I hadn't seen another way. I had to get my information together. Now it was time to put my cards on the table. In some ways he was the most important person in the story, and I didn't understand who he was. I told him a Robert Frost line Mike Basile had quoted to me—

We sit in a circle and suppose
The secret sits in the middle and knows—

I told him the different theories about him. First the anybody-can-snap theory from his old friend Gay Roberts. In his second year Dennis was isolated and unhappy, and one bad thing after another had happened, and finally he'd snapped, Gay said. Then the evil-genius theory that 'Akau'ola and Frank Bevacqua subscribed to, that Dennis had planned this to a fare-thee-well, even playing the governments off each other.

Dennis walked along beside me, now and then giving the faintest smile. He stared straight ahead through dark brown Armani glasses, but I got a glimpse of his eyes, deepset, dark, and big, liquid holes. His beard was shaved neatly around his mouth and cheeks, but his shoulders in the cutoff T-shirt were hairy, and he had a funny walk in his jeans shorts. His manner was so sensitive it sometimes seemed feminine. It was hard to say what he was thinking. Though he held up his hand when I talked about visiting the Gardners, didn't seem to want to hear that, and when I told him the Gay Roberts theory he went to a plate-glass window and traced a big circle on it with his finger and then stuck his finger in the middle of the circle.

I told him about the dream that had gotten me going. The night after I met Deb's mother and told her that I was writing this book, I was awakened by a dream. I was carrying two pairs of hiking boots away from an auction house where I had left them, to be sold to others. Now they were mine again, and filled with ice. Someone tried to stop me from taking them away, but I brought the boots back to my house and set them by the fireplace and set a big book to burning in the fireplace. The flames licked out and thawed the ice in the two pairs of hiking boots, and for the rest of the night I lay awake and felt that Deb Gardner's spirit was in the room pushing me forward. She was angry, she haunted this earth, she wanted me to do this job. That dream had propelled me for three years. The spiritual force of her untold story was bigger than any person's resistance, we were all being carried along by it.

I suppose it was a sign of Dennis's breadth of mind that he expressed no contempt when I said all that, seemed to accept the truth of it.

The first time we hit Union Square, he pulled out a folded piece of paper and read me a proposal. He said that I could convey his terms to my publisher, so I will report its fuzzy outline here: Dennis had no interest in this book coming out, but if I waited till 2007 to publish, he would tell me everything. He'd be 55 then— and I won't go into his explanation—but if I decided to publish before 2007 he'd never tell the story, and as it was I had very little. . . .

I should have anticipated the calculations, I should have thought of Emile in the jailhouse with the double-jeopardy proposal. But I didn't. I didn't laugh at his last-chance bluff or wave the New Zealand blood analyses at him—provided to me after Dennis's former attorney Clive Edwards, now the minister of police in Tonga, had authorized their release—and say he had no idea what I knew.

What I told Dennis was that he should come forward because of the havoc the case had left in many hearts, the idea that a person could kill a person and walk away from it. I reminded him of what Pastor Vili Vailea had told him outside the jail, ask forgiveness from the girl's family. I reminded him of the scene in *Crime and Punishment* where Raskolnikov goes to Sonya the prostitute and tells her what he's done and says, What should I do? and Sonya says, Go to the crossroads and kiss the ground in four directions and say I have sinned, and God will give you life again.

It was my own form of bluff. Wayne Gardner didn't want to talk to Dennis. If Dennis came to him on bended knee he'd know what to do with him and it wasn't talk. Wayne's good wishes had carried me far, but he wasn't along on this one, he didn't want Dennis forgiven, forgotten, forewarned. He wanted Dennis imprisoned, or hanged back in Tonga, he wanted justice. I didn't tell Dennis that.

Because what was his idea of justice anyway? He had those mental boxes with lids that Marie had described. He was too interior a person to believe in justice, too private, his imagination too crazy and elaborate. This is just something that had happened between a couple of people. "She deserved it," he had said in Tonga, and maybe he still believed that, and in that sense he seemed to me evil. He had treated the murder as a kind of accomplishment, not something to be regretted.

So many other parties had been at the table, holding cards, each with their own idea of how the game should end. Peace Corps/Washington, Peace Corps/Tonga. The State Department. The Royal Family. The chargé d'affaires. The NANEAP director. Dennis's Uncle the Judge. Judge Hill. The police ministry. The Crown law department. Dr. Lebensohn. Dr. Stojanovich. The friends of Dennis Priven. Sidney and Mimi Priven. (Alice and Wayne had never taken a seat.)

Dennis had won that game. He'd maintained a poker face for three months. He'd studied the table more closely than anyone and of course he'd staked more than anyone else. He'd almost killed himself a couple of times to get out of Hu'atolitoli. Emile had refused to play the double-jeopardy game, but others of his friends had helped. He'd made a kind of confession to the Bird Lady to gain admission back into the human family (and to ensure her presence?) but her loving expectation that he would get help meant nothing to him. John McMath gave him a Bible he had read thoroughly, and he told Dr. Stojanovich he was Deb's Jesus Christ—or allowed Stojanovich to say as much on the stand—then in the States had told Dr. Lebensohn that Deb had led him on and crushed him. Two different stories. And each was the key to its respective legal doorway.

It was pointless to cite a larger social good, he was too sociopathic. So I appealed to Dennis's grandiosity. I said that what he had pulled off in Tonga was actually a stunning addition to the annals of crime. There was a brilliance to it, a negative brilliance for sure, but the amazing thing is that the story was unsung. I was going to change that—didn't he want to help?

"OK, if I'm as smart as you say I am, then how come it's not me with the big house by the lake in Seattle?"

"You're as smart as Bill Gates, you just care about different things."

Our positions were too far apart. And though we hung together for more than three hours, feeling each other out, and even made a charade of trying to come to terms, it wasn't going to happen.

For years I'd been picturing this encounter and always with explosive scenes. Dennis would get furious, violent, he'd attack me. He did get angry a couple of times, and I had the underlying sense that he was deeply disturbed, but it was a friendly meeting. He was a free man in Soho. We were having a civilized chat. Not once did I scream at him, not once did I try and shake loose the monster of Ngele'ia for all the world to see. For some little time I had put my hatred of him and my rage aside. And I was angry at myself for that, it seemed disloyal to Deb. But then he had done the same for me.

We were done. We went back to Dean and Deluca and I got a bottle of juice and he got a lemonade. We walked south again, sipping from bottles. "I want to show you my pictures," I said with finality.

We sat on a rusted iron stoop on Grand Street and I showed him 100 or so of the images I'd collected. He flipped impassively through the pictures of Deb, broke down when he saw a picture of Paul Boucher, lost it for a few minutes, had to walk off down the street. The monster—only thinking about his own bloody life.

Then he carefully drew something from his knapsack that he'd brought along, a stiff card with a blue edge, his membership in the Royal Nuku'alofa Martini Club, the group founded by Lampshade Phil in 1975, signed by John Sheehan. So, there had been good parts of Dennis's Peace Corps experience he'd never been able to claim.

I put my pictures away and he put away his card from the Royal Nuku'alofa Martini Club. We got our knapsacks on, had a moment's small talk. Then Dennis headed toward Broadway, I headed toward Lafayette. He didn't look back, I'm sure of that. But then neither did I.

When the case had ended, when the minister had learned that Dennis went free, a story went out in Tonga that a lot of people heard: Dennis Priven is dead.

He stepped off the plane in the United States, the story went, and then someone from the girl's family, the girl's uncle or brother, came up and shot him then and there on the tarmac.

The story went around like wildfire. Dr. Lutui heard it, policemen heard it. Dennis's neighbor S'no heard it. The head of A3Z heard it, members of Parliament heard it, Dennis's former colleagues and students at Tupou High heard it. Deb's colleagues and students heard it. And they believed it, too. They wanted to believe it. The story satisfied some deep social understanding, that if somebody killed someone, it would catch up with him—he would die.

I could never determine where the rumor started, but I was sure all the same of the source: It was 'Akau'ola.

The case had ripped his country's social fabric, and he had done his best to repair it with a romantic tale. The minister had never been able to accept the here and now of incompetence and compromise, he believed in a perfectible world just over the horizon that could be found in the books he had in his study, and in his own imagination. And as for his belief in justice, 'Akau'ola would not give that up, not for Dennis Priven or all the powers of the United States government.

So Dennis had died on the tarmac. And now having met him, it seemed to me that maybe 'Akau'ola was right. Certainly something inside him had died, on the tarmac or in Ngele'ia that night, when he had gotten out his knife.

I avoided Ngele'ia. It was a sad place. But every visit I made to Tonga there would come a day when out of respect I bicycled over to Deb's *fale*.

There it was still, at the end of the rugby field and across from the bush, though urban sprawl had somewhat pacified the bush and later occupants had expanded the house. The *fale* had been added onto in two directions, but you could still see the original lines. There were glass louvers in the windows, and the place was completely wired. One night I saw the blue glow of a television set.

A Tongan teacher lived there with his family. A Tongan teacher lived in Emile's house, too.

There were no *pālangi* teachers at Tonga High or Tupou High now. My country had done exactly what the King had asked of us. We had provided smart eager young teachers for the high schools and gotten Tonga to the next stage. It was Deb's accomplishment. And Dennis's too.

Peace Corps is one of the noble achievements of postwar American society, it got a self-centered country to look beyond its borders. Anyone who questions that should go see Dave Wyler doing foreign aid work today in Nuku'alofa or Greg Brombach counseling inmates at Soledad Prison or Jon Lindborg working for the State Department at the largely evacuated embassy in Jakarta. But Peace Corps was never forthright with the public about risks to volunteers, and from that original error stemmed the effort to smother this case.

What a few self-important American officials did in the Priven case was indefensible. They manipulated the Tongan justice system to get the verdict they wanted. They lied to the King and Privy Council to free a vicious murderer, and they deceived the head of their parent agency about the case, they deceived the vice president. Paul Magid said Peace Corps had a legal duty to tell the Gardners that Dennis had walked—they misled Deb's family. And one former official told me there had been an order from above to do that. They had covered the case up.

They did so to preserve their own careers, to preserve the American presence in the South Pacific, to preserve the churchly image of the Peace Corps.

In doing so, they had also served the interests of a murderous criminal. Dennis wasn't just an ordinary man who had snapped, and he wasn't an evil genius either, he was both things. He was a brilliant madman allowed to stay too long in the wrong spot who had lost control and then manipulated everyone around him with coldness and creativity. 'Akau'ola had not seen his like before. Yet Dennis could never have pulled off what he did without the powers of the American government behind him.

And really, how well had they served him? Not well at all. Dead on the tarmac, as 'Akau'ola said. The officials had walked away from it whistling, but the outcome had created fear in Dennis's own family and denied him any opportunity for renewed life.

Because he had gotten away with it, and every step he took, he was compelled to preserve that achievement. He could put it away in the box, but the box never went away. Six months after our meeting, he changed his phone number again and he retired from Social Security, just before his fifty-first birthday. More cleverness, more calculation to get away with it, to escape the consequences of my book, the consequences of his actions. He was still on his bicycle, rushing away in the dark!

He would always be getting away with it. It was the only thing he would ever get to do. My government had condemned him to that. He was behind the bricks, plotting forever to continue to get away with it. And what sort of life was that? Deborah Ann Gardner woke him up at night, too. No one forgets his first foreign country.

A Note on Sources

This book is a reconstruction of a historical episode more than a quarter century after the fact. The murder of Deborah Gardner and trial and freeing of Dennis Priven, while described in a series of articles in the Tonga *Chronicle* and several dispatches prepared by the *Chronicle* staff for the Associated Press, are treated in no other published sources. No Peace Corps history mentions the case. Biographies of the King of Tonga and Gerald Ford and Henry Kissinger also do not refer to the case. Jan Worth's novel, *Nightblind*, in which she describes the murder has yet to be published.

My book draws on hundreds of interviews conducted over several years with participants in these events and on extensive research of documents and archives, some public, some personal. I also gathered a great number of photographs that contributed to this writing.

In this section I will list major areas of my reporting and cite sources.

I. VOLUNTEER LIFE

The descriptions of Tonga 16's staging in San Francisco, training in Nuku'alofa, and first year in country are drawn from interviews with members of the group. I was able to interview about 15 of the 33 members of Tonga 16 in person and exchange letters with another 4 or 5. They include Michael Bednar, Paul Boucher, Rebecca Brickson, Greg Brombach, Gary Buelow, Victor Casale, Judy Chovan, Kelly Downum, Avis Johnson, Kay Johnson, Dave Johnson, Jon Lindborg, Francis Lundy, Rick Powers, Ron Pummer, Roni Mellon, David Scharnhorst, Rob Schlachter, Mark Stiffler, and Christopher Van Bemmell. I was also able to interview several of the group's trainers: Rod Hooker, Barbara Dirks, Sione Maumau, and Tovo'Uele. I met with many members of Tonga 11, 14, 15, 17, and 18 as well.

My description of Deb Gardner's life in Tonga and before that is based on records and interviews. Her effects include a half dozen letters from Deb to her mother. Thanks to the Lakes High School Class of 1971 and Washington State University, I was able to alert alumni to my project and received letters that Deb wrote to friends in 1975 and 1976. Her letters to Richard Kim, Carol Rainey, and Ellen Canfield, describing her Peace Corps experience were invaluable. High school friends of Deb's including Bruce Alber, Frank Kemery, and Doug Dearth, were also helpful. And my interviews with Karen Hegtvedt, Ellen Canfield, Carol Rainey, Barbara Agnew, and others threw light on Deb's life.

Deb's activities are also mentioned in several personal archives. Victor Casale provided me with copies of all his letters and a short story, "The King's Grip." Avis Johnson, Jon Lindborg, Jan Worth, and Mark Stiffler read to me from their diaries. I had full access to the Tongan journals of Michael Braisted, 1973–1977; the letters of Judy Chovan, 1975–1977; and the tape recordings of Greg Brombach to Pete and Beverley Brombach, 1975–1976. Finally, Frank Bevacqua provided copies of his eulogy and poem about Deb from October, 1976. My rendition of his poem at the end of Chapter 14 is slightly truncated and is published with his permission.

In a few instances, I have rendered conversations between Deb Gardner and other volunteers (Frank Bevacqua, Mike Braisted, Judy Chovan, Vic Casale, Doug Jackson, and David Scharnhorst) after working closely with each former volunteer and gaining his or her approval of the truthfulness of the account.

Many of Deb's students also helped me. They include the Honorable Fielakepa, 'Ana Ma'u, Temaleti Pahulu, 'Emmelina Tuita, and 'Asinate Matangi. Former staff at Tonga High, including Rob Beaver, Russ Denne, Dan Haddock, Pensimani Tupouniua, and Talahiva Tonga, contributed to this account. As did the widow of Alan Gardiner, Christine Gardiner.

In re-creating the life of Dennis Priven, I have drawn on interviews with members of Tonga 14, notably Emile Hons, Tom Riddle, Harold Green, Kevin Holmes, the late Ralph (Lolo) Masi, Berl Parker, John Myers, Tim and Kathy Sullivan, and Barbara Wilson, who is identified in these pages as the Bird Lady.

Dennis's friends Philip English, John Sheehan, and Galen Martin helped me, as did John and Marie McMath, his former principal, Kalapoli Paongo, his former colleagues Matthew Abel, Merv Hamer, Mesui Saofi, and Ian Sullivan, and several students including Claude Tupou, Likio 'Atiola, and Keith Moala. For descriptions of Dennis's youth in Brooklyn, I conducted interviews with about a dozen men who knew Dennis before and after Tonga, in the neighborhood, and at Brooklyn College, and with a half dozen friends of his late parents. Several of these people expressed the desire to remain anonymous, and with a couple of exceptions I made it a rule to leave them anonymous in the text.

Tom Riddle has produced many writings about Tonga, including his "Reminiscence of the Death of Deborah Gardner," prompted by my research. Emile Hons shared with me all his writings home, as well as numerous personal papers (including *Taimi Totonu*, the Peace Corps newsletters, and his eulogy for Deb, which is mentioned in Chapter 14).

The richest archive I drew on was the "Letters to Her Family of Barbara Williams." Barbara, whose name is now Barbara Wilson, freely shared an archive of nearly 160,000 words. In many instances, when describing Barbara's thoughts from this time, I lifted language from her letters. The most extensive of these borrowings is the description of Barbara's leave-taking in Chapter 25, with her many observations of nature.

Gay Roberts also shared her lengthy journals from 1975 and 1976, and Ian Sullivan shared Gay's letters to him.

II. THE EVENTS OF OCTOBER 14, 1976

My description of the murder of Deborah Gardner by Dennis Priven is based on numerous eyewitness and documentary accounts.

I met and interviewed more than a dozen Tongans who were in Ngele'ia when Deb was killed. They include Pila Naufahu, Le'ota Naufahu, Edgar Cocker, Takitoa Taumoepeau, Sateki Finau, Lile Pasa, To'a Pasa, Viliami Pasa, Tongia Tupou, Sekona Tu'uta, Havili Sefesi, 'Aleki Ongosia, and Pila Mateialona.

These events are also described in the trial and police records, which I cite below.

The description of the emergency-room scene came from Taniela Lutui and

Losimani Kapukapu. For an understanding of Deborah's wounds, I relied on Dr. Lutui and also on Dr. Steve Bergmann, of Moscow, Idaho, an emergency physician who analyzed the various reports from Tongan sources.

A twenty-page archive of notes and analyses in the case is also maintained by the Institute of Environmental Science and Research (formerly the Departent of Scientific and Industrial Research) in New Zealand. Wayne Bedford and Keith Chisnall, forensics officers at the Institute, provided that file to me in 2002 following authorization from the Tongan Ministry of Police and the family of Deborah Gardner.

Dennis made many statements about that night to friends, to his lawyer Tomi Finau and the Peace Corps lawyer Paul Magid, and to psychiatrists Kosta Stojanovich and Zigmond Lebensohn. The statements to Finau are recorded in the National Archives files cited below. His statements to friends were related to me by Emile Hons, Gay Roberts, and Barbara Wilson. Barbara's letters home also contain some descriptions of Dennis's statements. Paul Zenker, then a close friend to Dennis, declined to be interviewed for this book but did agree to answer some written questions by mail. I interviewed both Dr. Stojanovich and Dr. Lebensohn, and also Dr. Clark Richardson, who accompanied Dennis back to the United States on January 13–14, 1977. And I met several times with Dennis's former attorney, W. Clive Edwards.

Laura Boucher did not respond to my letters. Her husband, Paul, declined to be interviewed but agreed to answer my questions by e-mail. Paul took exception to several assertions in this narrative. He said that he never discussed personal relationships with Dennis before the murder, the two were not close in that way. He understood that Dennis was interested in Deb and she wasn't interested in him, but no more. He has only a vague memory of Dennis's July dinner for Deb. "I was part of no group to 'protect' him, or sympathize with him," Paul states. He has no memory of Dennis's confidences in jail, though Dennis might have made them, he says, and he cannot remember gatherings at his house at which the case was discussed.

III. The Tongans

My understanding of the actions of the Tongans and expatriate workers in Nuku'alofa in the time surrounding this case draws on many journeys to that region over three years, interviews in all the island groups of Tonga (with the exception of the Niuas), and several trips to New Zealand and Australia as well as to Indonesia, Holland, and England.

Many of the Tongans' actions are documented. Transcripts of the preliminary hearing in the case, October 25–26, 1976, exist in the "Michael Basile Papers" and the "General Counsel Files," both cited below. Three separate handwritten lawyers' records exist of the trial, Dec. 13–23, 1976. Judge Hill's record of 110 pages is maintained by the Supreme Court, Kingdom of Tonga. Tevita Tupou's personal notebook for the trial, about seventy pages long, also contains an extensive record of Clive Edwards and Tomi Finau's questioning of witnesses. The fullest documentation of the trial was created by Christine Oldroyd (now Davis). In reconstructing the preliminary hearing and trial scenes in Chapters 19, 23, 24, and 25, I drew on numerous interviews with participants or spectators, among them Elizabeth Wood-Ellem, Clive Edwards, Paul Magid, Rick Nathanson, Barbara Williams, Claude Tupou, Frank Bevacqua, Jen Crawford, Christine Davis, Laki Niu, Kosta Stojanovich, Emile Hons, and Tevita Tupou. Undoubtedly my best source on the events of the trial was the translator Kaveinga Tu'itahi. His memory of climactic moments was invaluable.

As to Sinoveti Motuliki's testimony, I interviewed Sinoveti and three of her sons, Tevita, Polikapi, and Taniela. My understanding of Tongan concepts of mental illness was drawn in part from two papers by Dr. Mapa Puloka, "A Commonsense Perspective on Tongan Folk Hearing" and "Avanga, What Is It?" I also relied on "Illness and Cure in Tonga: Traditional and Modern Medical Practice," by Siosiane Bloomfield.

Police holdings in the case include a book of photographs, the Seahorse knife still in evidence, and the private journal of Siaosi Maeakafa Aleamotu'a, lent to me by his widow Lute Aleamotu'a.

Officers and former officers of the Tonga Police who assisted me include: Haini Tonga, Kuli Taulahi, Siaosi Tui, George Blake, Taniela Faletau, Sinilau Kolokihakaufisi, 'Aisea Naeata, Nauto 'Ata'ata, Siope Matapule, 'Opeti Prescott, Faiva Tu'ifua, Vungakoto Pa'akolo, Siliveni Tautua'a, Semisi Tapueluelu, and Moleni Taufa.

Police Minister 'Akau'ola was dead when I commenced this project. My description of his life and attitudes is based on interviews with his two brothers, the present 'Akau'ola (the former 'Inoke Faletau) and Sione Faletau, his two adopted daughters, and his close associates. My report of his response to this case is based chiefly on interviews with former secretary to government Taniela Tufui, George Blake, Sinilau Kolokihakaufisi, Haini Tonga, 'Eleni 'Aho, and other officers; and with Rick Nathanson and Merle Anders, two former volunteers who were friendly with the minister. Crown Prince Tupouto'a also related his exchanges with 'Akau'ola.

I re-created 'Akau'ola's imagining of the hanging, at the end of Chapter 26, based on a visit to Mo'unga Kula, an examination of the hanging tools used by 'Akau'ola, and descriptions of 'Akau'ola's conduct as a hangman at two hangings in 1977 and 1982. His brother Sione, a minister, detailed his late brother's mental preparations and anxieties surrounding the duty.

To describe the Privy Council meeting of January 7, 1977, I relied on police documents shared with me by the Ministry of Police, and accounts of the meeting from three participants: Baron Vaea, Langi Kavaliku, and Dr. Sione Tapa. His Majesty, the only other survivor of that meeting, declined to be interviewed.

For the social life of Tongans and expatriates in Nuku'alofa in 1976 I have drawn on numerous accounts, written and oral. Chief among written sources is the Tonga *Chronicle*, which is held in several libraries and personal collections, including the *Chronicle*'s library and the Palace Traditions Office. Also: Rick Nathanson's collection; the library of Liahona High School, a branch of the Mormon church; the library of the University of the South Pacific/Tonga Centre; and the Auckland University library. I have also relied on the "Diary of Elizabeth Wood-Ellem" and the "Diary of Vearl Ferdon," parts of which Elizabeth and Vearl shared with me.

My knowledge of the career of Henry Hill is based principally on interviews with two lawyers who worked with him, Toby Hooper, a barrister in the Middle Temple, London, and Harry Waalkens, a barrister in Auckland. Tongan Chief Justice Gordon Ward also helped me, as did former Peace Corps country director Preston McCrossen. The characterizations of Hill's decline are based on interviews with Tongan lawyers and with his former lover 'Ofa Manu Mateaki, in Nuku'alofa, and former house servant Tausili Taufa in Auckland.

My treatment of Tongan history is based on written sources. They include principally the most authoritative work on Tongan history, *Queen Sālote of Tonga: The Story of an Era 1900–1965*, by Elizabeth Wood-Ellem, as well as *Island Kingdom: Tonga Ancient and Modern*, by I. C. Campbell, *Shirley Baker and the King of Tonga*, by Noel Rutherford; the edition of William Mariner's *An Account of the Tongan Islands*, published by the Vava'u Press in Nuku'alofa, and *The King of Tonga*, a biography by Nelson Eustis. I have also relied on *The Environment of Tonga, A Geography Resource*, by Wendy Crane, and *Tongan Place Names*, by Edward Winslow Gifford (a bulletin of the Bishop Museum in Hawaii). Edgar Tu'inukuafe's *A Simplified Dictionary of Modern Tongan* was also helpful, as was *An Intensive Course in Tongan*, by Eric B. Shumway.

Finally, the most important published source I made use of was the *Tongan Dictionary* written by C. Maxwell Churchward, a wondrous resource.

IV. Washington

I hope and expect that my accounts of American officials' actions will be scrutinized. Almost every word attributed to an American official in this book can be documented. In no case did I take the liberty I describe above in recreating dialogue between volunteers. This includes Mary's vision of the "true killer" of Deb Gardner in Chapter 15 and the threat prepared by General Counsel Russ Lynch for the Gardner family, "Once Out, All Out," in Chapter 20.

Several archives contain records of the American officials' statements and actions.

The Government of Tonga provided two crucial documents to me, the letter of John Dellenback to the King of Tonga in February 1977, and the letter of the Priven parents to Prime Minister Tu'ipelehake in December 1976. Perhaps, most important, Government sources in 1979 released to Mike Basile the sixteen pages of Privy Council documents that I quote from in Chapter 26. Included in the Privy Council documents are Robert Flanegin's January 5, 1977, letter to Prince

Tu'ipelehake making promises to the Tongan government. (This same letter has been heavily redacted by my government.) The Police Ministry of Tonga also released to me a follow-up statement by Flanegin giving the exact character of Dennis's detention.

The American records of the case are far more voluminous than the Tongan records, though in many cases they are redacted.

I relied principally on six sets of records:

1. The 1989 Peace Corps Freedom of Information Act release to me. These 120 pages of documents include numerous notes created in the case by the Peace Corps Office of Special Services. I believe I may have the only copies of these records that still exist anywhere.

2. General Counsel's Files, the Dennis Priven Case, National Archives and Records Administration. Roughly 500 pages of documents in the case were released to me by the National Archives in 2001. While heavily redacted and repetitive, they include Paul Magid's handwritten notes from his trip to Tonga in October 1976, cable traffic between Peace Corps/Tonga and Peace Corps/Washington, and many internal memoranda created in this highly unusual case.

3. Peace Corps Directors' Files, the National Archives. These include many of John Dellenback's notes and memoranda in the case, as well as records left by his successor, Carolyn R. Payton.

4. Gerald R. Ford Library records pertaining to Michael Balzano, John Dellenback, Mary George, and John Stiles.

5. The Armistead I. Selden Jr. Congressional Collection papers maintained at the University of Alabama, W.S. Hoole Special Collections Library. This voluminous collection of Selden's files while Ambassador to New Zealand, Tonga, Fiji, and Samoa includes numerous references to the Tonga case, including cables and letters.

6. The Personal Archive of Michael Basile. As troubles developed in the Peace Corps office in Tonga in 1976, Basile, who is now a scholar of international studies, created a record of his interactions with Mary George and others. Notable among these is a chronicle of events surrounding the murder and a

report on Mary's interactions with volunteers. Basile's papers also contain letters that Basile received relating to the case and Basile's own, "Three Letters to Guy Gattis," in 1997.

As I say, these are my principal sources. I also received numerous records from the Peace Corps in response to several Freedom of Information Act requests between 2000 and 2003, a small sheaf of records from the State Department in 1989 and 2001 after FOIA requests, and records from the Social Security Administration in 2003.

It would seem helpful to enumerate how I learned about several officials in this book.

MARY GEORGE

Mary George declined to help me with this book. I conducted interviews about Mary with many former volunteers and officials. The officials include Richard Hailer, Frank Guzzetta, Paul Magid, Dorothy Rayburn, Preston McCrossen, John Dellenback, and Nancy Blanks-Bisson. I interviewed some friends of Mary's, including George Fulcher, Simon Dobbs, Robin Judge, and Dan Tufui. And countless former volunteers offered reports of Mary's actions. Among the most important of these was Mary's meeting with Deb on October 14, 1976. This is the only instance in the book in which I recreate dialogue without having access to at least one of the participants in the meeting. Deb related parts of this exchange to several volunteers, including Mark Stiffler, Frank Bevacqua, Francis Lundy, John Myers, and Dave Wyler. Mary related it to Mike Basile. I have relied on these second-hand accounts.

Mary also left a documentary trail. A résumé and application are held at the Ford Library. Several of her own letters are included in the Selden papers. She is described in letters from Clark Richardson to Mike Basile, from Vickie Redpath to Kelly Downum, from Barbara Wilson, formerly Barbara Williams (the Bird Lady), to her parents, and by Judy Chovan to her parents, among others. Questions about Mary's suitability that arose in Washington are documented in Avis Johnson's diary, April–May 1976.

Mary's vision of a Tongan killer is reported in detail by Mike Basile in his chronicle of that period. It is also documented (though both officials heard about it secondhand) in Preston McCrossen's letter to Carolyn Payton, Peace Corps director, 1978, and Clark Richardson's letter to Payton, 1978 (both these letters are in the National Archives). Mary's encounter with Siaosi Maekafa Aleamotu'a that night, October 17, 1976, is documented in Aleamotu'a's diary for the year 1976. Two other visions Mary experienced are described in Basile's chronicle and in a record of the July 1977 meeting by Richard Hailer with unhappy volunteers in Nuku'alofa, a record that Basile also retains.

I conducted interviews with several Peace Corps staffers who served under Mary: Roger Bowen, Clark Richardson, Makaleta Lovo Fifita, Viliami Kapukapu, Moala Funaki, Pila Mateialona, Mike Basile, and Dave Wyler.

MICHAEL L. BASILE

Mike Basile obviously was one of my principal sources in writing this book. I have drawn extensively on his papers and many days of interviews over more than three years of association. His "Letters to Guy Gattis," 1997, were invaluable. In describing Mike's thoughts in Chapter 10, as he learns about the murder and begins a search, and in Chapter 15 as he looks for Mary on October 17, then finds her and meets with her and takes her home, I have lifted several vivid phrases from the Gattis letters.

JACK ANDREWS

Andrews's cables and phone conversation with John Dellenback, October 1976, are documented in the National Archives collection of the Peace Corps director, above. The two letters questioning the nature of Andrews' relationship to Mary George, written by Richardson and McCrossen, are also contained in that collection.

I conducted interviews with many former Peace Corps country directors in the NANEAP region under Andrews, including Terry Marshall, Nancy Blanks-Bisson, Dan Cantor, James Scanlon, Jack Burgess, Richard Hailer, Preston McCrossen, Jimmie White, Robert Graulich, Allan Gall, and Everett Woodman. Numerous former Peace Corps officials also contributed to my understanding.

John Dellenback

Again, I did interviews with former headquarters staff, including Mike Balzano, Fred Hansen, Vito Stagliano, Gretchen Handwerger, Jack Burgess, Jack Daniels, Dottie Rayburn, Paul Magid, Gerry Vitulano, and David Searles. The crucial call to Alice Gardner on October 14, 1976, was described to me by Alice, by Justine Corby Southall, and at Peace Corps, by the former secretary Libby Videnieks.

Dellenback's campaign trip on October 15, 1976, was covered in the *Connecticut Daily Campus*, the University of Connecticut newspaper, retained at the University of Connecticut Library.

Russ Lynch

Lynch's comments in the case are documented by himself and others in the National Archives General Counsel's files. Lynch's comments to Special Services officer Bette Burke are documented in the 1989 release to me. Lynch's close friend and associate, Jack Ganley, also provided information.

Robert Flanegin

The chargé d'affaire's cables are included in State Department releases to me, in Peace Corps Director's files at the National Archives, and in the Selden papers in Alabama. Letters by Flanegin are also included in the Selden papers. Flanegin's close friend, the late Philip Vandivier, was of assistance, as was his former associate, Harlan Lee.

Armistead I. Selden Jr.

The Selden papers offer a clear picture of his service and his visits to Tonga. The October 1976 visit in Chapter 8 is recorded by many former volunteers, including Greg Brombach, Rick Nathanson, and Mike Basile. Emile Hons, George Fulcher, Rick Nathanson, and Dave Wyler were interviewed about that visit.

V. Peace Corps History

In discussing the history of the Peace Corps I have made wide use of books, congressional records and interviews. In this connection, I had some access to the library at Peace Corps headquarters, including files related to Peace Corps deaths.

The best book on the history of the Peace Corps in this era is *The Peace Corps Experience: Challenge and Change*, by P. David Searles. Other important Peace Corps books include *A Moment in History: the First Ten Years of the Peace Corps*, by Brent Ashabraner; *Come as You Are: the Peace Corps Story*, by Coates Redmond; and *What You Can Do For Your Country: an Oral History of the Peace Corps*, by Karen Schwarz.

I have relied on back issues of *RPCV Writers & Readers*, a periodical edited by John Coyne, the writer and former Peace Corps volunteer. Coyne also shared an unpublished article, "Notes on the Founding of the Peace Corps." My descriptions of the Michelmore case in Chapter 18 come from Ashabraner and Redmond's books.

Former Peace Corps officials who helped me to understand the political culture of the Peace Corps in the 60s and 70s include Warren W. Wiggins, William Josephson, Jack Burgess, Tom Scanlon, John Keaton, David Searles, Victor Basile, and Chuck Hobby, among others. My description of the Peace Corps psychological screening history is based on numerous sources, principally interviews with former volunteers. Searles has a fine section about this in his book. This policy is also treated in the National Archives Director's files and in Congressional testimony by Sargent Shriver in 1962 and 1963 and by John Dellenback in 1975 and 1976 and Sam Brown in 1977.

My treatment of the Kinsey case in Chapter 18 is based in great measure on a large file on the matter retained by the National Archives. John Coyne wrote an article about the case for *RPCV Writers & Readers*.

VI. Miscellaneous Events

For my description of the actions of writers and journalists in this case, I have turned to libraries and writers. Reporters who had some knowledge of the matter who have assisted me include Peter O'Loughlin, Svein Gilje, Robert Keith-Reid, Tavake Fusimalohi, Siosaia Fonua, Rick Nathanson, the Honorable Tu'isoso (Paua Manu'atu), and Charles Salzberg. The Fiji National Archive in Suva provided access to two articles in the *Fiji Times* treating the case. The writer Joseph

Theroux has shared his fictional writing about the case, "Island Fever," an unpublished story written in 1979. The two Internet accounts I quote from in Chapter 31 are a sermon by Craig Moro and "Protect the Handicapped," a short story by Ingrid Richardson.

The novel that Bill Kinsey quotes in his journal in Chapter 18 is Wright Morris's *Ceremony in Lone Tree*. The song of praise that Marie McMath quotes in Chapter 31 was written by Gloria and William J. Gaither.

I have made glancing reference to the case of Thalia Fortescue Massie, the Navy wife whose false rape claim delayed Hawaiian statehood, in Chapter 19. I drew on the book, *Rape in Paradise*, by Theon Wells, and from discussions with Joseph Theroux.

Acknowledgments

M any people helped make this book.

David Hirshey wrote his first letter about Tonga for me sixteen years ago as a magazine editor, then when he became a book editor advised me, in the most genteel publishing tradition, to "get the #$%@!! out there and find the %!@#$ twist." In the past two years, David has read many, many drafts of this work, always pushing me to remember the !(@#$% story. Assistant editor Nick Trautwein brought a deadly ear for the false note. Susan Weinberg, HarperCollins's publisher, trusted me enough to tell me the worst.

I am thankful to my agent, Joy Harris, for her patience, and to two editor friends for support. Richard Stengel of *Time* wrote a persuasive letter of sponsorship when I first went to the Kingdom, and Peter Kaplan gave me a lifeline: He told me to move to the South Pacific and continue to write for his newspaper, the *New York Observer*. I am grateful to Peter's readers and his boss, Arthur Carter.

My mother, Ellen Weiss, and mother-in-law, Barbara Kling, provided invaluable comment on portions of my manuscript. The writers Daniel T. Max and John Paul Newport also saw early drafts of this book and laid on their hands.

Many others provided timely assistance: Wahab Ali, Steve Bergmann, Rev. Carl Bosteels, John Chapman Chester, Alta Clabaugh, Lynn Darling, Cindy Ducich, Vearl Ferdon, Linda Fontana, Ray Fruchter, Dana Gappa, Steve and Melissa Gregor, Nina George Hacker, Jan Higgins, Jeff Kellogg, Ellen Ketcham, Beth Kling, Anton 0. Kris, the late Zigmond Lebensohn, Harlan Lee, Michael Lewis, Tupou Lindborg, Adam Liptak, Rebecca Liss, Michael Massing, Kevin

McKeown, David M. Sachs, Charles Salzberg, Justine Southall, Kosta Sto-janovich, Jack Sullivan, Jonathan W. Tweedy, Rob Walker, Eve Weiss, and Eve Williams. The writer John Spritz got me to the Pacific in the first place, while Grant W. Fine, a lawyer and investigator, Beth Silfin, a lawyer, and Gregory Lomas, a writer, saved me from panic on a number of occasions.

I am indebted to librarians around the world, particularly Natalie Mahony at Auckland University Library, archivist Donna Lehman at the Gerald Ford Library in Ann Arbor, Brian Kamens at the Tacoma Public Library, Terri Goldich at the Dodd Center at the University of Connecticut, and Milton Gustafson and Beth Lipford at the National Archives and Records Administration.

The long list of former Peace Corps volunteers who assisted me begins with those who loved Deb Gardner. Emile Hons was generous and brave company, from Marin to Ma'ufanga. Judy Chovan led me into a girl's heart. Frank Bevac-qua's honesty and anger were a resource through years of mistakes, as was Mike Braisted's forbearance. Victor Casale was angrier than anyone at first, then gave me a brother's loyalty. (His parents, Fiore and Michelina, treated me like family, too.) Mark Stiffler and Jon Lindborg read to me from their diaries and showed me the solid friendship that Deb had also drawn upon. David Scharnhorst is a private person, but he shared the most personal stories out of the same impulse that caused his old buddy Kelsey Downum to do so, too—the feeling that a departed friend would want them to.

Many other former Peace Corps volunteers and officials also helped me. Among the volunteers: Merle Anders, David Arnold (editor of the newsletter 3/1/61), Roger Bowen, Bill Brandewie, Rebecca Brickson, Charles Contant, Denny DiPaolo, Dennis Ferman, Jack Fones, Bob Forbes, Bill Goldsmith, Harold Green, Don Greer, Merv Hamer, Julianne Hickey, Kevin Holmes, Rod Hooker, Doug Jackson, Avis Johnson, Kay Johnson, Dave Johnson, Lorraine Leiser, the late Ralph Masi, Al and Dorothy Massinger, Macon McCrossen, John Myers, Ted Nelson, Lori Osmundsen, Berl Parker, Rick Powers, Ron Pummer, Richard Stoll, Bill and Barbara Stults, Tim and Kathy Sullivan, Chris Van Bemmel, Nathaniel Winstanley, Wayne Witzell, and Tami Somerville. Bruce McKenzie told me the legend, and Fred DuBose got me going in '88. Caroline Wulzen gave me moral clarity when I didn't have it. Francis Lundy was the straightest shooter I've ever met. John Sheehan came and went with cutting insights.

The officials include Michael Balzano, Victor Basile, Nancy Blanks-Bisson, Jack Daniels, Brad Dude, Jack Ganley, Teodoro I. Guambana, Richard Hailer, Fred Hansen, Gretchen Handwerger, John Keaton, Paul Magid, the late Bob Martin, Galen Martin, Preston McCrossen, Charlie Peters, Clark Richardson, Jim Scanlon, Tom Scanlon, David Searles, Ed Slevin, Vito Stagliano, Libby Videnieks, Gerry Vitulano, and Jimmie White.

Help should not be confused with endorsement. Some Peace Corps people were leery of my work and maybe still are, but they showed me the civility of guiding someone with whom they had real differences. On that note, Bill MacIntyre was one of many who declined to participate in this project—but in doing so, he shared his concerns, which were instructive. Paul Boucher also differed with me, but helped me out of respect for the Gardner parents' wishes and the historical character of the case

Two legendary Peace Corps founders understood exactly what I was doing and gave of their time and records. Warren W. Wiggins and William Josephson have my gratitude. A latter-day legend in Peace Corps administration, Jack Burgess, met with me on several occasions and also read an early draft, providing important emendations.

Several former-volunteer writers assisted me. That list must begin with Jan Worth; if not for her hand at the start, I wouldn't be finishing up now. She showed me true grace—pressing me to do this book even as she was working on related material. ("Mine's better," she explained.) Rick Nathanson provided Dickensian commentary on volunteer life; Tom Riddle's comments were more in the Burroughs vein. The conservator of Peace Corps legend, John Coyne, schooled me, while Joe Theroux, the conservator of Polynesian legend, took me from Bligh to Brando, and "Rain" to Rainbow Warrior.

For their help in understanding Deb Gardner's school days, I am thankful to Bruce Alber, Ellen Canfield, Pat Caraher, Doug Dearth, Mary Aiken Fishback, Jane Hedges, Karen Hegtvedt, Joe Jensen, Richard Kim, Debbie Laughlin, and Kerry and Steve Lawson. The memories of two of Deb's closest friends, Barbara

Agnew and Frank Kemery, were essential to me, as were the letters that Carol Horan Rainey had preserved.

I drew on many volunteers' personal archives, notably those of Emile Hons, Vic Casale, and Judy Chovan. Five other collections were essential.

Early on in my work, Mike Basile handed me a box of documents and diaries from 1976 and 1977, which I read that night, and again on many more nights over the following years, with awe. Mike allowed me to call on his judgment on countless occasions, while his wife, Jan, grounded me.

Two other volunteer archives helped me establish the sequence of events. Out of love for Deb Gardner, Greg Brombach gave me free rein in the fifty or sixty tapes of his Peace Corps experience in 1976 and 1977 (and his parents, Pete and Beverly, welcomed me into their home several times). Gay Roberts Devlin, a former Volunteer Serving Abroad in New Zealand, allowed me to copy her faithful and epigrammatic diaries.

The last two archives were humbling gifts to this work. Mike Braisted overcame his customary reserve to grant me unlimited access to his poetic, reflective, and at times joyous journal. He transformed this book. Barbara Wilson, formerly Barbara Williams (the Bird Lady), painstakingly transcribed her voluminous letters home from Tonga in 1975 and 1976 (Tanya Bray helped) then handed me the disk. The elegance, rigor, and detail of these writings were models as I tried to negotiate the same wrenching terrain. Barbara never fully approved of my point of view, but acted from pure intellectual motivation: the desire to share important knowledge.

Present-day volunteers also helped me, notably David Dauer, Chuck Hayes, and Holden Warren. Tony Paquin set up my lecture at the University of the South Pacific Centre in late 2002, at which Anna Urban criticized my work in a way that markedly changed it. She has my gratitude.

In at least one respect, I hewed to Peace Corps tradition, for I got a lot more from foreigners than they got from me.

In England, I had the assistance of Christine Davis, Simon Dobbs, and Toby Hooper; in Holland, Jen Crawford; in Fiji, 'Epeli Hau'ofa, Robert Keith-Reid, and Steven Vete.

In Australia, I thank John and Ann Connan, Paul Cotton, Phil and Yvonne Doherty, Carson and Pam Flint, George and Pat Fulcher, Joan Furby, John and Yvonne Hepplewhite, 'Ofa Malupo, John, Marie and Phil McMath, Peter O'Loughlin, Laureen Outtrim, Ian Sullivan, and the leading *pālangi* scholar of Tongan society, Elizabeth Wood-Ellem.

In New Zealand, the Institute of Environmental Science and Research Limited (and its officers Wayne Chisnall and Keith Bedford) played a vital role in this project: It preserved and then released forensic analyses of evidence. Philip English was the best company for an American in 1976 and again a quarter century later. Donald K. Hunn, New Zealand's High Commissioner to Tonga, 1976–1979, and his wife, Janine, put me up in Otaki Gorge, and put up with my enthusiasms. Other Kiwis who helped me include Matthew Abel, Rob and Judith Beaver, Russ and Joyce Denne, Christine Gardiner, Dan and Linda Haddock, Robin Judge, Taniela Lutui, Judith Macfarlane, the Motuliki family, Le'ota Naufahu, To'a Pasa, and Harry Waalkens.

I thank the people of Tonga. I came to the Kingdom a stranger in March 2001. The Tongans heard me out and accepted me. Without them, this book would be a shadow of itself.

Sela and 'Atolo Tu'inukuafe made their home mine for months, and the Tu'inukuafe children treated me as a sibling. Across town, Dave and 'Anau Wyler fed me, lodged me, gossiped with me, and corrected my Tongan. I think of Dave and Kitione Mokofisi as my brothers. They and my other fellows at the Nuku'alofa Club restored my spirits on countless nights in Nuku'alofa.

Tevita Tupou and Clive Edwards have often been at odds in Tongan public life going back to the Priven case, but their differences were of no account in this project. Both men gave me time and records.

Numerous officers in Clive's command, beginning with Lau'aitu Tupouniua, provided assistance. Others include Haini Tonga, 'Opeti Prescott, Kuli Taulahi, Nauto 'Ata'tata, 'Aisea Manu Naeata, Taniela Faletau, Semisi Fifita, and John W. Racine. Moleni Taufa showed me repeated hospitality at Hu'atolitoli.

The family of the late 'Akau'ola—the present 'Akau'ola, Sione Faletau, Sokopeti Faletau, and Lautoa Faletau—helped me to understand the Minister.

Kaveinga Tu'itahi, the former court interpreter, enriched this story in any number of ways. Talahiva Fine Tonga met with me often out of love for her old colleague. Dr. Mapa Puloka shared his writings, songs, and spirit. The Supreme Court Registrar Manakovi Temaleti Pahulu provided me with numerous records, while the Hon. Tu'isoso (Paua Manu'atu) offered me free use of the archives of the Tonga *Chronicle*. On countless occasions, Makaleta Lovo Fifita shared her memories of Peace Corps.

I am grateful to the Crown Prince Tupouto'a, and also to the Honorable Fielakepa. I thank Baron Vaea and the Hon. Ma'afu. Also the Hon. Langi Kavaliku, Dr. Sione Tapa, Taniela Tufui, Judge Gordon Ward, and the Hon. Tu'a Taumoepeau Tupou.

Others in Tonga who assisted me include 'Eleni 'Aho, the Aleamotu'a family, Likio 'Atiola, Feleti 'Atiola, Paula Bloomfield, Edgar Cocker, June Egan, Sepuloni Faupula, Tuna Fielakepa, Kaimana Fielakepa, Sateki Finau, Steve Finau, Toe'umu Fineanganofo, Papiloa Foliaki, Moala Funaki, 'Ahongalu Fusimalohi, 'Eseni Helu, Sister Keiti, 'Ofa Koloi, Viliami Kapukapu Lavulo, Losimani Kapukapu, Paul Karalus, 'Asinate Matangi, Pila Mateialona, 'Ofa Manu Mateaki, 'Ana Ma'u, Pilima Misa, Siaosi Moengangongo, Laki Niu, Kalapoli Paongo, Meleana Puloka, Dr. Tilitili Puloka, Carl Riechelmann, Cindy Soakai, Lata Soakai, Claude Tupou, Vili Vailea, Adiloa and Naran Prema, Mesui Saafi, Litia Simpson, Tangaloa (Chris Cross), the late Palavilala Tapueluelu, 'Aisake Takau, Tausili Taufa, Takitoa Taumoepeau, the men of the Tonga Club, Mahe'uli'uli Tupouniua, Lu'isa Vivili, Saane Tupou, 'Amelia and Sekona Tu'uta, and Bill Vea.

Apologies to all whose names I've forgotten; it's been a long haul. And though I wish I could name the many Brooklyn residents who helped me in researching Dennis's life, most desired anonymity. I am grateful to an anonymous friend for scriptural interpretation.

Throughout this project, I leaned on four friends. Dave Merandy was my companion in the woods; Dan Swanson (the author James North) got me out of them. Vusenga Helu gave me faith in bars, and churches, too, while John Averitt offered me kinship at the best and worst times.

Finally, there are the people from whom I asked the most. I told Wayne and

Alice, and John and Barbara and Craig, too, that I believe some good may come of this. I only hope that time proves me right. My wife, Cynthia Kling, is the wisest person I know, and maybe the wiliest, too. She multiplied this book.

Koloa pē 'a e Tonga ko e fakamālō—Tongans say the only wealth is to be thankful. All those who were so helpful when I so needed help have my deepest thanks. *Mālō au pito!*

Insights,
Interviews
& More . . .

*

About the author

About the book

Read on

Meet **Philip Weiss**

Linda Fontana

PHILIP WEISS has been a contributing
writer to the *New York Times Magazine*
and a contributing editor to *Esquire*
and *Harper's* magazine. His first novel,
Cock-a-Doodle-Doo, was published by
Farrar, Straus & Giroux in 1995. ❧

How I Came to Write
American Taboo

I'M AWARE of how much suffering this story records, and how much pain my work caused. But Wayne Gardner feels halfway redeemed in his relationship with his daughter. He says he likes to go hunting now and not kill the birds, just watch them light on the pond.

In writing about Deb Gardner, I'd served my own interests, but also those of others. I had helped the people of Tonga achieve recognition for the things they did on Deb's behalf. I'd helped a lot of Americans and Polynesians replace a disturbing oral legend with solid facts, facts that their societies could deal with. The politicization of Deb's murder was a piece of history that had been covered up. Now that it wasn't, there was a sense that my country might even do something about it. That is what Deb Gardner would have wanted. And though it appears that the official unraveling is going to take a while, the story had been a long time in coming out.

Telling the story lifted a burden for a lot of volunteers who had to live with this thing, with the uneasy feeling that it was not strictly private, as the Peace Corps had thuggishly insisted it was. Some of these folks are thankful to me for telling the story with depth and dignity. Emile

> **"** The politicization of Deb's murder was a piece of history that had been covered up. Now that it wasn't, there was a sense that my country might even do something about it. That is what Deb Gardner would have wanted. **"**

How I Came to Write *American Taboo*
(*continued*)

Hons had trouble talking about this episode at the start. By the end he was choking up on national television and giving viewers an understanding of the moral discomfort of an old injustice. David Scharnhorst wrote me to say that though he'd read the book, he didn't need to. He knew it already, and anyway, it made him feel that he should have done more for Deb at the time. David still thought about the fact that if he'd only stayed in Tonga, and taught at Tonga High, maybe.... I reminded David that he'd done a lot for Deb by explaining who she was for people who were learning about her years after her death. The story required someone outside it to tell it. I feel that I was of service to those folks.

That makes it easier for me to be open about the pleasure I got from this work. I felt so lucky to be on this case, and I loved every minute of it. It is embarrassing to admit, when you are talking about a tragedy, but it's true. The usual cacophony of doubt about what I was up to subsided, and I cried and whooped through the whole process. Chapter 12 ends with the statement that Wayne Gardner wants to help—that crucial moment in the story when, after a year and a half, the Gardner family got in touch with me. Well, in an earlier draft I said that when I got the news I went running through the streets of Auckland, crying and shouting with joy, and then ran back to my

> " The story required someone outside it to tell it. "

hotel to call Frank Bevacqua to give him the news.

This was the best material a writer could ask for, and no one had told it before, so I had complete conceptual freedom. When Mike Basile told me about Mary George's vision in the cathedral, when Tongan policemen told me about 'Akau'ola's pity for Deb Gardner, I knew I was in the territory of Somerset Maugham and Robert Louis Stevenson. And I did my best to rise to the material.

It was hard to move on to the next thing. Then one day in Tonga I realized it was over. We had just pulled up across from Vaiola Hospital when Sara Ely Hulse, CBS's associate producer, got a phone call. She spoke for a few seconds and then, snapping the phone shut, fist-pumped the air. "Yesss!"

A former Peace Corps volunteer from Texas had sent Sara an old Super 8 film from Tonga that included a few moments of Deb Gardner at the dance at the Dateline Hotel, five days before she died. I'd been trying to get that film from the Texas volunteer for several years, but had never convinced him. Now national TV had sprung it loose. Later, back in New York, I'd see the few seconds on that film of Deb walking across the dance floor. She was statuesque and her features were strong and clean, like her dad's. Her hair flooded down over her white shoulderless dress. But right then in Tonga, I sensed that this story now belonged to others. It was out there. Other people were going to explore it in their own ways. ▶

> Right then in Tonga, I sensed that this story now belonged to others. It was out there. Other people were going to explore it in their own ways.

How I Came to Write *American Taboo*
(continued)

What do you do when a four-year chapter of your life ends? I was trying to throw myself into something else, but my mother said it didn't have the passion of the Deb Gardner story. What story would? I could wait the rest of my life and never find another story with that same passion. And anyway, I'd never be able again to do what I had done to uncover Deb's story, go to as many insanely out of the way places, for so long, to track things down, to talk to anyone who had heard her voice or touched her hand before she died. It was a youthful mission. ∽

> **My mother said it didn't have the passion of the Deb Gardner story. What story would?**

American Taboo
The Aftermath

I had promised to give Dennis Priven a copy of my book when it came out, and so I mailed one to him in June 2004. A few days later the package came back to me with a big white sticker pasted over the address on which was printed the word "refused" in a formal serif font. Later I learned that similar notices were appended to returned letters others sent to Dennis urging him to come forward. I could imagine Dennis sitting in front of his computer for an hour to dream up just the right expression of his feelings about our packages, then acquiring just the right white sticker. Then the drama of bringing the thing back to the post office, or merely dropping it in a box. He would have had to handle it for some time, even if he did not open it.

When the book was published, the CBS News show *48 Hours* picked up the story, and I camped out for two days in a van with tinted windows as the production team made preparations to photograph him. He was still living at his parents' former apartment in Sheepshead Bay, and on day two he came out in his jeans shorts and T-shirt. "That's him," I said. Like other wide-shouldered burrowing creatures, he blinkingly examined the skies before returning to his apartment to adjust his wardrobe for the conditions. Then he stepped back out into the world, inside his detached blue bubble, and was only

> " A few days later the package came back to me with a big white sticker pasted over the address on which was printed the word 'refused' in a formal serif font. "

briefly irritated when Susan Spencer, the correspondent, tried to talk to him, before going on his way.

The story had taken on a life of its own. In October 2004, Norm Dicks, the congressman from Deb's hometown, called on the U.S. Attorney's Office in Seattle to explore the possibility of criminally charging Priven. Wayne Gardner repeated his call for official action in the case, joined by others who had read my book or read about it. They wanted to see Dennis charged with the crime, to see the Peace Corps and the State Department held to account for their actions in the case; they wanted a full investigation, during which officials who had played the Tongan government to allow Dennis to get away with it would put their right hand on a Bible and testify. My country's diplomatic adventures in Iraq and across the Middle East contained echoes of the Tongan model. Foreign opposition, foreign laws and voices, were to be countered with arrogance and indifference.

The Peace Corps was still covering it up. Twenty-eight years after Deb's death, they asked to make a statement to CBS. A badgerish executive named Patrick J. Hogan said repeatedly that the Peace Corps had no records surrounding the case, so it couldn't really respond to the charges, apart from saying that it felt awful for Deb's family. What a load. I had tons of records I could show them. I could bring them to the five-hundred-page file in the nearby National Archives,

> 66 Patrick J. Hogan said repeatedly that the Peace Corps had no records surrounding the case, so it couldn't really respond to the charges, apart from saying that it felt awful for Deb's family. What a load. I had tons of records I could show them. 99

where the Peace Corps surely would have a better hope than I had of determining what lay under the blackouts in the documents.

The Peace Corps could at the very least have looked at its active computer file on Dennis Priven, in which it was stated that he had a Completion of Service certification, with no qualification. No asterisk saying, Oh, he murdered a fellow volunteer. It could have made some determination about whether that was appropriate. Instead, as the Tacoma *News Tribune* reported in the months after my book appeared, they came up with a hollow set of "talking points" to respond to the case without exploring it. The Peace Corps' concern was the same as it had been in 1976: it didn't want to look bad.

The agency was having a bad year. Congress had launched an investigation following a series of articles in the *Dayton Daily News* about assaults on volunteers. It was the same story as the Deb Gardner case—a volunteer got into trouble in a foreign land, and the Peace Corps' instinct was to protect the program even if that meant forgetting about the individual.

Volunteer culture somehow fostered this denial. It was the pride of volunteers to say, We are the good guys. We don't carry guns, we don't invade and occupy other countries, and darn it, we have a hard time getting attention for all the good we do. Why are you even talking about this case?

I went to a conference of returned Peace Corps volunteers in Chicago a month or so after my book came out, and the kumbaya ▶

> " It was the same story as the Deb Gardner case—a volunteer got into trouble in a foreign land, and the Peace Corps' instinct was to protect the program. "

feeling was everywhere. Peter Yarrow sang songs till midnight and was made an honorary returned Peace Corps volunteer, and though the banquet offered a minute of silence for dead Peace Corps volunteers (thereby acknowledging the presence in the room of Chelsea Mack and her mother Donna, who are now trying to establish a monument in Washington to fallen Peace Corps volunteers, a monument that would include the name of Chelsea's brother, Jeremiah Mack, who died in Niger), the atmosphere of self-congratulation in the hall, especially at a time of gruesome news from Iraq, was overwhelming. Returned volunteers were asked to call out "hello" in the language they had learned overseas, and one after another they did so. No one spoke in Tongan, and I was tempted to call out "*malo e lelei.*" But I stayed silent at the back of the room.

No one in that community was demanding answers about Deborah Gardner. The freeing of her murderer was a curiosity, irrelevant. It took Brian Finley, an attorney working for a public agency in Chicago, to come forward (not in his official capacity but out of compassion for Deb) and interrogate the Peace Corps in sharp letters to its spokesman.

"It appears you can say that 'I don't know what they [former officials] knew' because you and the Peace Corps have chosen to not look for the information or interview key people in the incident. Or, worse, you do know the answers but choose to continue

> 66 Returned volunteers were asked to call out 'hello' in the language they had learned overseas, and one after another they did so. No one spoke in Tongan, and I was tempted to call out '*malo e lelei.*' But I stayed silent at the back of the room. 99

to rub salt into the wounds of Deborah Gardner's family by claiming ignorance. The Peace Corps' conduct in this matter was scandalous at best, corrupt and illegal at worst. And your statements are irresponsible and still smell of a cover up thirty years later."

The Peace Corp wrote Finley back in a mollifying tone. It didn't work. He responded: "Your claim that 'the Peace Corps mourns [Deborah Gardner's] life and every loss of a volunteer's life' rings hollow. Can you explain how helping Dennis Priven get away with murder is 'mourning' Deborah Gardner's life? What else has the Peace Corps done to 'mourn' her life besides misrepresent to the Tongan government and the Gardner family that Dennis Priven was going to be committed to a psychiatric hospital when he returned to the United States?"

I loved the Peace Corps. I said it at any public appearance. It was a noble institution. Yet in betraying Deb Gardner, and continuing to betray her by refusing to open up the matter, it had revealed the bureaucratic and self-righteous cold spot in its heart: loving humanity and singing kumbaya all night, but not caring about a real person. Volunteers and staff could rationalize their indifference to Deb's story however they wanted, blaming the messenger, saying that this story was sensational and I was a commercial author. Still there was that body lying on the beach that they had done nothing about for almost thirty years now. Soldiers would have done a lot more. ▶

> Can you explain how helping Dennis Priven get away with murder is 'mourning' Deborah Gardner's life?

After the book came out, I heard from a lot of former volunteers. One former Tongan volunteer wrote to say he didn't think Dennis would have done what he did if he and Deb had been in the States. I agree.

The mystery in my book was always the violent enigma of Dennis Priven. Was he carried away on a wave of passion or emotion? Or did he calculate the thing coldly? Only at the end of this project did I come to my own solid understanding that he had calculated it. He was a sociopath, a detached personality who needed to operate on his own terms, utterly, and in 1976 these terms included the need to kill someone.

Any feeling I still had that Dennis had been insane under the legal definition of the term ended when I went with CBS to Tonga, my tenth trip to the country. One day Sara Ely Hulse, the associate producer, and I met a former student of Dennis's, Tevita Tupou, who was now a math teacher at Tonga College. "Was Dennis insane?" I asked.

Tupou shook his head. "I do not think so. That Thursday in Chemistry class, he handed out notes for the rest of the term and said that he might be going away. We did not understand what he meant 'til the next day, when we learned he had killed the girl."

It was more evidence that Dennis had been planning murder. Step by step, he had followed out his disturbed logic, beginning with the assertion "She deserves it," on to bicycling to Deborah Gardner's hut with a backpack full of weapons.

> 66 [Dennis] was a sociopath, a detached personality who needed to operate on his own terms, utterly, and in 1976 these terms included the need to kill someone. 99

I believe Dennis's calculations before killing Deb went something like this: the rules didn't apply to him; he was more intelligent than anyone else and understood that the rules were based not on principle but on social efficacy—a means of organizing human conduct under a hypocritical shell of religiosity and fairness; the only idea that had any weight for him was that Deb Gardner had "wronged" him and some others. She had used her womanly powers to turn the men around her into slavering swine and therefore was a fitting victim, in his disturbed view. He would demonstrate his superiority by carrying out Deb's elimination.

People in Tonga liked to say that Dennis was a genius. Certainly, his mind was far ranging enough to forecast a number of outcomes of the murder. He had examined Mary George, the State Department, the Tongan government, and the Peace Corps in the same way that a demolitions expert will study concrete piers in a stadium before blowing it down. He had judged that when the blow came all these entities would see little interest in hanging him by the neck until he was dead. Rather, they would see to it that he was set back on his feet on the shores of the United States. As he was.

My book ends with a spiritual idea: that no one is an island, that despite his best efforts, Dennis is wracked by conscience, that the murder of Deb Gardner will forever haunt him. I had some proof of this. During the events in Tonga, Dennis needed, and called upon, the understanding of close friends; ▶

> 66 My book ends with a spiritual idea: that no one is an island, that despite his best efforts, Dennis is wracked by conscience. 99

he needed a community. And he seemed still to need one. After Emile Hons made comments about Dennis in the press in California that were both compassionate and judgmental, someone, perhaps a friend of Dennis, sent Emile an anonymous e-mail to the effect that Dennis had received psychiatric treatment after his return to the United States. This friend of Dennis seemed to think the public should know this.

So I thought that Dennis had some community; and I still hoped that he would come forward—as I hoped that official investigations of the case might loose the words of officials and former volunteers who still held their secrets about the case. All these people had sent me, in one way or another, a giant "refused" sticker. They were waiting for a priest or a congressman before they talked, and I sure hope they get one. Deb deserves it, and so does her family and the country.

This book represented a coming-of-age for me. In my mid-forties, this project brought me back to my own starting point as a global citizen—the South Pacific—and having finished the book, I have moved on to other Pacific stories involving cultural collisions. So the Peace Corps had achieved its goals with me. I had grown up in a privileged American household, but after my meetings with volunteers in Samoa in 1978, the Peace Corps had gotten me to look outside my borders, maybe for the rest of my life.

> I hoped that official investigations of the case might loose the words of officials and former volunteers who still held their secrets about the case. All these people had sent me, in one way or another, a giant 'refused' sticker.

If You Loved This,
You'll Like . . .

White Mischief (1982)
by James Fox

The classic account of an officially "unsolved" murder among British expatriates living in Kenya's Happy Valley in the late 1930s and early 1940s, Fox's portrayal of a micro-society of decadent royals carousing and scheming in the African jungle is endlessly fascinating, and his tense murder mystery structure grabs the reader from the start and never lets go.

In the South Seas (1896)
by Robert Louis Stevenson

Robert Louis Stevenson's timeless narrative of his sailing journey through the South Pacific is one of the earliest depictions of the collision of European and Polynesian cultures, as well as an epic adventure every bit as thrilling as the best of his fiction.

In Cold Blood (1965)
by Truman Capote

Widely credited with launching the "nonfiction novel" genre (a blend of journalistic subject matter and literary style and structure), *In Cold Blood* is the harrowing story of the brutal murder of a rural Kansas family at the hands of two shiftless drifters. Capote's smooth, precise prose coupled with a tale riveting in its macabre intensity makes for an unmissable read. ▶

A Dream of Islands (1980)
by Gavan Daws

Subtitled "Voyages of Self-Discovery in the South Seas," Daws's overlooked classic is a brilliant investigation of five famous figures of the nineteenth century—Paul Gauguin, Robert Louis Stevenson, Herman Melville, John Williams, and Walter Murray Gibson—who were in some way shaped or defined by their often astonishing South Pacific experiences.

Crime and Punishment (1866)
by Fyodor Dostoyevsky

Some consider this to be the greatest novel ever written, but at the very least it is an infinitely complex, totally engrossing journey inside the mind and soul of a tortured, hapless murderer. The novel's dark, claustrophobic atmosphere and ingenious plotting make it the prototypical literary thriller.

The Ugly American (1958)
by William J. Lederer and Eugene Burdick

A remarkably prescient look at the massive missteps and blunders in American policy and diplomacy in Southeast Asia, this important worked prompted an entire rethink of the attitude and stance of the United States in the region. Through a mix of anecdotes, analysis, and detailed research, *The Ugly American* is a work that demanded change, and the echoes of its influence can still be felt today.

Don't miss the next book by your favorite author. Sign up now for AuthorTracker by visiting www.AuthorTracker.com.